Introduction to Beam Dynamics in High-Energy Electron Storage Rings

Introduction to Beam Dynamics in High-Energy Electron Storage Rings

Andrzej Wolski
University of Liverpool, UK

Morgan & Claypool Publishers

Rights & Permissions
To obtain permission to re-use copyrighted material from Morgan & Claypool Publishers, please contact info@morganclaypool.com.

ISBN   978-1-6817-4989-1 (ebook)
ISBN   978-1-6817-4986-0 (print)
ISBN   978-1-6817-4987-7 (mobi)

DOI   10.1088/978-1-6817-4989-1

Version: 20180601

IOP Concise Physics
ISSN 2053-2571 (online)
ISSN 2054-7307 (print)

A Morgan & Claypool publication as part of IOP Concise Physics
Published by Morgan & Claypool Publishers, 1210 Fifth Avenue, Suite 250, San Rafael, CA, 94901, USA

IOP Publishing, Temple Circus, Temple Way, Bristol BS1 6HG, UK

Contents

Preface   viii

Acknowledgements   x

About the author   xi

1   Introduction   1-1
1.1   A brief history of electron storage rings and their uses   1-1
1.2   General features and subsystems   1-7
       1.2.1 Magnets   1-8
       1.2.2 Radiofrequency cavities   1-10
       1.2.3 Feedback systems   1-13
       1.2.4 Vacuum systems   1-14
       1.2.5 Diagnostics   1-17
       1.2.6 Control systems   1-20
       1.2.7 Injection systems   1-20
       1.2.8 Personnel protection   1-22
1.3   Some examples of electron storage rings   1-23
       References   1-27

2   Linear optics   2-1
2.1   Co-ordinate system and transfer matrices   2-1
       2.1.1 Drift spaces   2-4
       2.1.2 Dipole magnets   2-5
       2.1.3 Quadrupole magnets   2-8
       2.1.4 Radiofrequency cavities   2-11
       2.1.5 Transfer matrices in three degrees of freedom   2-13
       2.1.6 Fringe fields and edge focusing in dipole magnets   2-16
2.2   Betatron oscillations   2-18
       2.2.1 Hill's equation and the Courant–Snyder parameters   2-18
       2.2.2 The matched distribution in a periodic lattice   2-20
       2.2.3 Betatron phase advance and the betatron tunes   2-22
       2.2.4 Action–angle variables   2-23
2.3   Emittance   2-24
2.4   The closed orbit   2-26
2.5   Dispersion   2-27
2.6   Coupling   2-29

2.7   Momentum compaction factor   2-31

2.8   Synchrotron oscillations and phase stability   2-34

       References   2-36

3   Synchrotron radiation   **3-1**

3.1   Features of synchrotron radiation   3-1

       3.1.1 Radiation power spectrum   3-2

       3.1.2 Brightness   3-4

       3.1.3 Opening angle of the radiation beam   3-5

       3.1.4 Polarisation   3-6

3.2   Radiation damping and quantum excitation   3-7

       3.2.1 Damping of synchrotron oscillations   3-8

       3.2.2 Damping of betatron oscillations   3-10

       3.2.3 Quantum excitation   3-13

3.3   Natural emittance and lattice design   3-17

       3.3.1 FODO lattice   3-18

       3.3.2 Double-bend achromats   3-20

       3.3.3 Theoretical minimum emittance lattice and multi-bend   3-21
              achromats

       References   3-23

4   Nonlinear dynamics   **4-1**

4.1   Chromaticity   4-1

       4.1.1 Natural chromaticity in a storage ring   4-2

       4.1.2 Correction of chromaticity using sextupole magnets   4-3

       4.1.3 Coupling and nonlinear effects from sextupole magnets   4-5

4.2   Resonances   4-6

4.3   Dynamic aperture   4-10

4.4   Energy acceptance   4-12

       References   4-14

5   Collective effects   **5-1**

5.1   Touschek scattering and space charge   5-2

5.2   Ion trapping   5-5

5.3   Wake fields, wake functions, and impedances   5-6

5.4   Potential-well distortion   5-11

5.5   Microwave instability   5-13

        5.5.1   Microwave instability in a 'cold' beam   5-15

        5.5.2   Energy spread and beam stability: Landau damping   5-15

5.6   Coupled-bunch instabilities   5-17

        References   5-22

6   Further topics   **6-1**

6.1   Advanced tools for beam dynamics   6-1

6.2   Some other phenomena   6-4

        6.2.1   Spin dynamics   6-5

        6.2.2   Beam–beam effects   6-7

        6.2.3   Electron-cloud effects   6-10

6.3   The future: 'ultimate' storage rings, and beyond   6-12

        References   6-15

Preface

Electron storage rings lie at the heart of some astonishingly productive scientific research facilities. Particle colliders have made possible discoveries that have changed our understanding of the fundamental laws of nature. Synchrotron light sources have provided insights into chemical and biological structures and processes, which impact our lives through applications in medicine, technology, industry, and the environment. Light sources are even used for research in art and archaeology. Electron storage rings are also a fascinating subject of study in their own right: the design and construction of a modern storage ring requires a deep understanding of a wide range of physical phenomena, and the application of some of the most advanced engineering and technical systems.

The goal of this book is to describe some of the main features and principles of operation of modern electron storage rings. The scale and complexity of these machines means that this is a challenging task, especially if it must be accomplished in a concise form. Inevitably, I have had to make a number of compromises in setting the scope of the book and the depth to which the material is developed. I have tried to focus on those aspects that, in my experience, are really essential for some understanding and appreciation of how electron storage rings work. I have also attempted to produce a text that will be of value for a rather diverse audience, including those who have a general interest in the accelerators used for so much important scientific research, scientists involved in research making use of electron storage rings and who wish to know more about the machines on which their work depends, and students who come across storage rings as a topic in their under-graduate or postgraduate studies.

As an inevitable consequence of the numerous choices I have had to make in the selection and presentation of the material, some chapters are more technical than others. Chapter 1 outlines the history of electron storage rings and describes in largely qualitative terms the purpose and features of some of the principal technical subsystems. This chapter will, I hope, be accessible and comprehensible to a general audience. Chapters 2 and 3 develop the beam dynamics in more technical detail, including some of the mathematical ideas and techniques that form an essential part of the accelerator physicist's toolkit. I have avoided detailed derivations of key formulas and results, giving references instead to numerous books that develop the material in more detail and with greater rigour. However, I believe I am not alone in finding that a certain amount of maths often helps to clarify ideas that can otherwise seem rather vague and abstract. Chapters 4 and 5 tackle some of the more challenging (but rather important) aspects of beam behaviour in storage rings, namely nonlinear dynamics, collective effects, and beam instabilities. Although I do present some important formulae in these chapters, the relevant phenomena and the underlying physical mechanisms are described largely in qualitative terms. Finally, chapter 6 mentions, rather too briefly, some further interesting aspects of electron storage rings and outlines the current direction of accelerator research motivated by the various applications for which storage rings have been used.

My hope is that whatever the nature of your interest, you will find at least some parts of the book of value. For myself, even after working nearly 20 years in the field of accelerators, I am constantly astonished (and pleased) by how much I still have to learn about these elegant and fascinating machines.

Acknowledgements

I would like to thank my colleagues and students at the Cockcroft Institute, Daresbury Laboratory, and at the University of Liverpool for endless support and advice, and in particular for many useful, enlightening and enjoyable discussions on accelerators, on physics, and on learning and teaching.

About the author

Andy Wolski

Andy Wolski is a Professor of Accelerator Science in the Department of Physics at the University of Liverpool in the UK, and a member of the Cockcroft Institute of Accelerator Science and Technology. After obtaining a PhD in theoretical physics from the University of Manchester, he trained as a teacher and taught science in secondary schools for a number of years, before returning to physics research at Daresbury Laboratory in 1998. Two years later, he moved to the Center for Beam Physics at Lawrence Berkeley National Laboratory in the USA, before coming home to the UK to take up his present post at the University of Liverpool. His accelerator research (theoretical and experimental) has related to a variety of projects, including synchrotron light sources, linear collider damping rings, and a number of circular colliders. As well as being active in research, he has contributed to numerous accelerator schools by delivering lectures on a range of topics in beam dynamics. Andy was awarded the 2017 IOP Particle Accelerators and Beams Group Prize for his contributions to understanding the dynamics of particle beams at high-energy accelerators.

Introduction to Beam Dynamics in High-Energy Electron Storage Rings

Andrzej Wolski

Chapter 1

Introduction

1.1 A brief history of electron storage rings and their uses

In the late nineteenth century, experiments with cathode ray tubes led to the discovery of the electron. The components of a cathode ray tube perform all the essential functions of the components in a modern accelerator and include: a particle source, some means of acceleration, components for controlling the trajectory of particles, and a way of detecting or observing the particles. By modern standards, however, the components in a cathode ray tube are extremely simple. A heated wire at a negative potential (the cathode) is the source of electrons, which are accelerated (in a vacuum) by a static electric field towards a fixed anode. The electrons may be steered using electric fields perpendicular to the accelerating field, or by a magnetic field generated using loops of wire carrying electric currents. The beam of electrons is observed either by the fluorescence of residual gas within the tube, or by the light produced when electrons strike a coating of fluorescent paint on the inner surface of the tube. In the next section of this chapter, we shall outline the corresponding systems in electron storage rings that are being constructed today. It is worth noting that nearly 30 years elapsed between the development of the first cathode ray tubes by William Crookes and others, around 1869, and the discovery of the electron by J J Thomson in 1897.

Accelerators were developed continually through the twentieth century [1], motivated by the desire to understand the physical properties of materials, and the laws of nature, at ever more fundamental levels. This required increases in beam energy and intensity, and improvements in the quality and stability of the beams being accelerated. One way to achieve a high energy is to use a sequence of electric fields. The strength of a static electric field is limited by the fact that any material will break down once the field reaches a certain value. However, by using oscillating fields, the ultimate energy is effectively limited only by the length of the accelerator.

doi:10.1088/978-1-6817-4989-1ch1
© Morgan & Claypool Publishers 2018

In a drift tube linac for example, particles pass through a sequence of conducting tubes. At any given moment, any two adjacent tubes are at potentials of opposite polarity: this means that the maximum potential can be kept below the breakdown limit. If the potentials of the tubes (and hence the fields between the tubes) are static, then particles are alternately accelerated and decelerated as they pass successive gaps, and there is no overall acceleration. However, if the potentials (and the fields) oscillate in synchronism with the particle motion, then it is possible to arrange for particles always to see accelerating fields, and for their energy then to increase over the entire length of the linac. Although linacs are still widely used today, they have a drawback in that the higher the energy required, the longer the accelerator has to be.

The invention of the cyclotron by Ernest O Lawrence in 1934 was a major step forward for achieving high energy: the key feature of a cyclotron is the use of a magnetic field to bend the particle trajectory into a spiral. This means that particles can be accelerated multiple times by an oscillating electric field, in the same way as in a linac; but the size of the accelerator is kept small by 'reusing' the same field. However, the energy that can be reached by particles in a cyclotron is limited by the size and strength of the magnets needed to maintain the spiral trajectory. Relativistic effects also lead to technical challenges in cyclotrons once particles reach energies at which they move at speeds close to the speed of light. Although cyclotrons are capable of accelerating particles to energies of several GeV, the magnets required are extremely large and can weigh tens or hundreds of tonnes. Nevertheless, cyclotrons are still widely used today for accelerating protons or ions (with mass larger than protons) for nuclear physics experiments, or for radiotherapy. However, electrons have a much lower mass than protons, so reach relativistic speeds at much lower energies, and this makes cyclotrons unsuitable for producing high-energy electron beams. As the energy of an electron increases, the radius of its trajectory in a magnetic field increases; however, its speed approaches a limit, the speed of light. Therefore, in a cyclotron, electrons start to take longer and longer to complete each turn of the spiral trajectory as their energy increases, until eventually their paths are no longer synchronised with the oscillating electric field needed to produce the acceleration.

Already in the first half of the twentieth century, it was realised that the limitations from relativity on electron beam energy could be overcome if particles could be kept on a circular, rather than spiral, trajectory as they were accelerated. This required the strengths of the magnetic fields controlling the trajectory to be increased in proportion to the momentum of the particles; conveniently, increasing the magnetic field created (by electromagnetic induction) an electric field that would accelerate the particles. With a suitable geometry, including some shaping of the poles of the magnet to control the field strength as a function of radius, a machine could be constructed that accelerated particles and at the same time confined them within a toroidal vacuum chamber. The resulting accelerator was known as a betatron. Although the concept dates back to 1922, the first successful machine was not demonstrated until 1940. Using betatrons, it became possible to accelerate electrons to tens of MeV; but the concept was soon superseded by the synchrotron. In a synchrotron, the magnetic fields used to guide the electrons are increased with beam energy, in the same way as in a betatron. However, rather than using electromagnetic induction to accelerate the particles, a synchrotron uses oscillating electric fields, in a similar way to a cyclotron. To compensate for changes in the speed of the electrons as they increase in energy, the oscillation frequency of the accelerating field is varied. In other words, the accelerating field oscillation is synchronised with the revolution frequency of the particles travelling around the ring: hence, the name 'synchrotron'. The first synchrotron was constructed (from a modified betatron) in the late 1940s, at the Woolwich Arsenal Research Laboratory in the UK. A second machine, purpose-built by the General Electric Company at Schenectady, New York, USA, followed soon after, and achieved an electron beam energy of 300 MeV.

After the first electron synchrotrons, a number of proton synchrotrons were developed with the aim of achieving beam energies of several GeV, motivated by the production of beams of high-energy particles for studies in nuclear physics. Synchrotrons were used to raise the energy of a beam injected at relatively low energy to the energy needed for an experiment, at which point the beam was extracted from the synchrotron and directed towards a target. However, it was soon realised that there were applications for high-energy beams (of protons or electrons) stored in synchrotrons for long periods. In principle, to operate the synchrotron as a storage ring all that was needed was to maintain the magnetic field strengths at constant values once the desired energy was reached. The use of synchrotron storage rings made it possible to perform colliding beam experiments, in which the trajectories of two beams of equal energy intersect at one or more points in the ring. This provides a significant advantage over fixed-target experiments, in which much of the energy of the incident particle is converted to kinetic energy of the collision products, and is therefore not available for generating new states of the interacting particles. When two particles collide with equal and opposite momenta, on the other hand, the total momentum is zero so all the energy is available for producing new states.

Colliders began to be developed in the 1960s. The first machine to be completed, AdA (Anello di Accumulazione, figure 1.1) was an electron–positron collider constructed in Frascati, Italy in 1961, by a team led by Bruno Touschek. The storage ring had a diameter of about 1.3 m and stored beams with energy 250 MeV. An electron–electron collider followed shortly after: VEP1 was operational by 1965 at the Institute of Nuclear Physics[1] in Novosibirsk, Russia, and achieved beam energies of 160 MeV in two storage rings each of 86 cm diameter. Numerous colliders have since been built, with each successive machine aiming for higher energy and/or luminosity. Machines have been constructed for colliding different combinations of lepton (electron or positron) and hadron (proton, antiproton, or ion) beams.

It has been known since the late nineteenth century that any charged particle will radiate energy in the form of electromagnetic waves when undergoing a change in

---

[1] INP is now named the Budker Institute of Nuclear Physics (BINP) in honour of its first director, Gersh Budker. Budker led the development of VEP1.

Figure 1.1. AdA (Anello di Accumulazione), the first electron–positron collider. AdA had a diameter of about 1.3 m, and stored beams with energy 250 MeV. It was constructed in Frascati, Italy, in 1961.

speed or direction. It was therefore expected, even before the first electron synchrotrons were constructed, that electrons in the magnetic fields in a synchrotron would emit radiation as a result of following a curved trajectory. The radiation from a beam of relativistic electrons is known as *synchrotron radiation* [2], and was observed for the first time in 1946, when light was seen emerging from the glass vacuum chamber of the General Electric Company synchrotron in Schenectady. Since synchrotron radiation limits the particle energy that can be achieved in an electron synchrotron, there were a number of studies to investigate its properties, and to confirm the theoretical predictions; the experience gained in handling synchrotron radiation for these studies opened the way for its use in a wide range of research. The value of synchrotron radiation lay in the intensity of the light that could be produced, compared to conventional sources, especially at short wave-lengths. This enabled experiments to be performed that had previously been difficult or impossible, notably into the properties of materials (for example, auto-ionization in gases). Despite a growing community of researchers primarily interested in synchrotron radiation as a tool for scientific investigation, the use of synchrotron radiation was for many years 'parasitic' on machines built for high-energy physics; it was not until 1968 that Tantalus, a 240 MeV electron storage ring at the University of Wisconsin, became the first machine constructed specifically for the production of synchrotron radiation [3].

In the decades following the construction of Tantalus, the use of synchrotron radiation for scientific research became more widely established [4]. Many new facilities were built, and advances in technology enabled significant improvements in the brightness, stability, and spectral range of the synchrotron radiation that could be provided for users. Conventionally, synchrotron light sources are classified in four 'generations'. First-generation light sources consist of electron storage rings constructed primarily for high-energy physics applications (i.e. colliders), with the synchrotron radiation produced by the dipole magnets being used parasitically. Although the colliding beams are tightly focused at the interaction point to generate high luminosity, the beam size in the rest of the ring tends to be relatively large: this limits the brightness of the synchrotron radiation produced from the electron beams in the dipole magnets. Brightness is a measure of the radiation intensity per unit area of the source, and is an important figure of merit for many light source users. Second-generation light sources are those (such as Tantalus) constructed specifically for the production of synchrotron radiation, but with the radiation coming principally from the dipole magnets used to steer the beam around the storage ring. The beam size in a second-generation light source can be smaller than that in a collider, but the brightness of the synchrotron radiation is still low by modern standards.

Advances in the design of the magnetic lattice in electron storage rings led to the development of third-generation light sources. In particular, it was found from detailed studies of the effects of synchrotron radiation on the electron beam producing the radiation that lattice designs were possible, allowing an improvement in brightness by several orders of magnitude (see section 3.3). At the same time, magnets were developed specifically for generating synchrotron radiation with desirable properties for light source users. These *insertion devices* (undulators and wigglers [2, 5]) consist of sequences of short dipole magnets of alternating polarity; an example of an undulator, in the Advanced Photon Source at Argonne National Laboratory in the USA, is shown in figure 1.2. A key feature of third-generation synchrotron light sources is that the storage ring is designed to optimise the production of high-brightness, short-wavelength radiation from insertion devices (although the radiation from the dipole magnets is often used as well).

Particles passing through dipoles and insertion devices in storage rings generally produce radiation independently of other particles in the beam, so that the intensity of the radiation is proportional to the number of particles. However, if the size of a bunch of particles is small compared to the wavelength of the radiation being produced, then the particles can radiate coherently, i.e. acting as effectively a single particle. The intensity of the radiation in that case is proportional to the square of the number of particles. Since the number of particles is potentially very large (of order 10^9 or more), the intensity of coherent synchrotron radiation can be many orders of magnitude larger than for incoherent radiation. Since bunches in an electron storage ring are typically several millimetres in length, any coherent radiation from an entire bunch has a large wavelength and does not propagate efficiently through the vacuum chamber. However, it is possible in some circumstances for substructures to develop within a bunch, leading to coherent radiation at wavelengths below a millimetre. A *free electron laser* (FEL [6–9]) is an accelerator designed to produce coherent synchrotron radiation, either from an electron storage ring, or from a linear accelerator delivering the beam for a long undulator. FELs are sometimes known as fourth-generation light sources. Short-wavelength (extreme ultra-violet or x-ray) FELs are based on linacs rather than storage rings, and have a

Figure 1.2. The canted undulator for sector 23 at the Advanced Photon Source. Sector 23 is operated by the General Medicine and Cancer Institutes Collaborative Access Team (GM/CA-CAT), a part of the Biosciences Division at Argonne National Laboratory. The GM/CA CAT has been established by the National Institutes of Health's National Institute of General Medical Sciences and National Cancer Institute to build and operate a national user facility for crystallographic structure determination of biological macromolecules by x-ray diffraction. (With permission from: Argonne National Laboratory.)

further advantage, in addition to the intensity of the radiation, in being capable of producing extremely short pulses of radiation, in some cases of order of a picosecond (10^{-12} s) or less. However, linac-based light sources are generally limited to serving only a few (perhaps three or four) users at any one time; third-generation light sources, however, provide much greater capacity. Depending on the size of the storage ring, a third-generation light source may serve several dozen user beamlines at once.

Developments in accelerator technology since the first-generation light sources have been matched by advances in the user beamlines: instruments designed for high-intensity, short wavelength radiation are required to steer and focus the beams of synchrotron radiation, to select particular wavelengths and polarisations, and to detect the radiation after interaction with a sample. The combination of accelerator and radiation beamline technology now available enables third-generation synchrotron light sources to support a highly diverse range of scientific research. The radiation can be used in various different ways [10–12]. For example, imaging can be performed in much the same way as in medical x-ray imaging, but the intensity of synchrotron radiation makes it possible to resolve much finer structures. The ability to select and combine images at different specific wavelengths also makes possible techniques that can provide images of structures on the scale of a few billionths of a metre. The ability to produce and select radiation at specific wavelengths from infra-red to x-rays also enables spectroscopy techniques based on how different materials absorb or reflect light of different wavelengths. Detailed spectroscopic measurements can provide important information on the structure and properties of materials with applications in information technology, engineering, biology, and medicine. Finally, diffraction and scattering experiments can reveal the atomic structure of materials ranging from minerals and ceramics to polymers and proteins. Particularly in the life sciences, the ability to determine the structure of highly complex molecules has provided deep insights into numerous biological processes, often with important medical consequences (for example, in understanding how particular medicines work). The value of synchrotron light sources for fields as diverse as materials science, information technology, engineering, biology, and medicine is such that about fifty facilities are currently in operation around the world, with several new facilities either proposed, under construction or being commissioned. Electron storage rings seem certain to continue to play an important role in scientific research for many years to come.

1.2 General features and subsystems

The detailed design of a synchrotron storage ring will depend on the intended application, but the overall structure generally follows an established pattern [13]. The beam travels within a vacuum chamber in the form of a tube with an internal aperture of (typically) a few centimetres. The tube is bent in (roughly) the shape of a circle with a circumference that could be anything from a few hundred metres to several kilometres. Particles are guided and focused using magnets placed at intervals around the ring. Electrons travelling through magnetic fields lose energy by radiation (in the case of relativistic particles, this is termed *synchrotron radiation*); to replace the lost energy, radiofrequency (RF) cavities are used to provide an electric field that accelerates the particles on each revolution. The electric fields in the cavities oscillate at frequencies of (typically) a few hundred megahertz; the frequency of this oscillation must be synchronised with the revolution frequency of particles around the ring, so that a particle will arrive at a cavity at approximately the same phase of the electric field oscillation on every turn. In this way, the energy of each particle in the storage ring remains roughly constant. In electron storage rings, the particle energy is typically in the range from a few hundred MeV, to many GeV.

Understanding the principles of a synchrotron storage ring and understanding much of the beam dynamics requires some knowledge of the magnets and the RF cavities. However, storage rings rely for their operation on many other types of component; altogether, a storage ring will usually be constructed from several thousands, or tens of thousands, of separate components. Some familiarity with the different types of components needed for a storage ring is often helpful in understanding even some of the apparently more academic aspects of beam dynamics. This is especially true in the context of technical limits on beam parameters.

Components of all different types in a storage ring must, to some extent, work together; but for convenience, they are grouped into various subsystems. Thus, the magnets, perhaps with their power supplies, cables, and cooling systems, will form one subsystem; the RF cavities, with the RF power source, waveguides (for transporting the RF power to the cavities) and electronics for controlling the amplitude and frequency of the fields in the cavities, will form another subsystem. Further subsystems include the vacuum system, diagnostics, feedback systems, control system, injection system, and personnel protection system. In this section, we outline the principal components within each subsystem, and the roles that they play in a storage ring.

1.2.1 Magnets

Magnetic fields are used to guide particles through the vacuum chamber, and to control the size (i.e. the cross-sectional area) of the beam as it moves around the ring. Dipole magnets produce a uniform vertical field that deflects the particles in the beam horizontally, so that the beam follows a defined path enclosed by the vacuum chamber. Quadrupole magnets (see, for example, figure 1.3) produce a magnetic field that varies linearly with distance from a path through the centre of the magnet: this type of field acts as a 'lens', providing the means to focus and control the size of the beam.

Other kinds of magnet are used for more refined control over the beam properties. For example, particles in the beam will have some (small) variation in energy, with the result that the focal length in a given quadrupole magnet will be different for different particles. This effect, termed *chromaticity* (see section 4.1), can be compensated by sextupole magnets, in which the field varies as the square of the distance from a line through the centre of the magnet.

Figure 1.3. The electron storage ring of the Advanced Light Source, Lawrence Berkeley National Laboratory. Dipole magnets (painted blue) steer the beam around the ring, while quadrupole magnets (red) provide focusing to control the beam size. Image courtesy of the Advanced Light Source, by SPat, CC BY-SA 3.0, https://commons.wikimedia.org/w/index.php?curid=25940137.

Most of the magnets in a storage ring will provide static fields; that is, the fields will be constant in time. The fields within these magnets must satisfy Maxwell's equations for magnetostatic fields[2]. Within the vacuum chamber, the current density (neglecting the beam itself) will be zero. This means that within the vacuum chamber, the magnetic field must have zero divergence and curl:

$$\nabla \cdot \vec{B} = 0, \quad \text{and} \quad \nabla \times \vec{B} = 0. \tag{1.1}$$

These equations are satisfied by fields with Cartesian components:

$$B_x = C_n r^n \sin(n\theta), \quad B_y = C_n r^n \cos(n\theta), \quad B_z = 0, \tag{1.2}$$

where C_n is a constant, r is the radial distance from the z-axis (along which the beam travels), and θ is the polar angle from the horizontal, transverse x-axis. Fields that can be expressed in the above form (1.2) are known as *multipole fields*. A field with $n = 0$ is a dipole field, $n = 1$ gives a quadrupole field, $n = 2$ a sextupole field, and so on. In terms of Cartesian co-ordinates, a general multipole field (with arbitrary n) is most easily expressed using complex notation:

$$B_y + iB_x = C_n r^n e^{in\theta} = C_n(x + iy)^n, \quad B_z = 0. \tag{1.3}$$

In a dipole, $B_x = 0$ and $B_y = C_0$. In a quadrupole, $B_x = C_1 y$ and $B_y = C_1 x$; and in a sextupole, $B_x = 2C_2 xy$ and $B_y = C_2 (x^2 - y^2)$. In practice, field strengths in accelerator magnets can vary quite widely. The maximum achievable fields depend on the type of magnet, for example whether the conductors in an electromagnet are normal or superconducting, or on the type of material used in a magnet constructed from permanent magnetic material. Normal-conducting electromagnets are perhaps the most widely used type of magnet in electron storage rings; dipole magnets of this type typically achieve field strengths of order 1 T, and quadrupole magnets can achieve around 0.8 T at the aperture limit set by the pole tips.
Solutions to Maxwell's equations can be constructed by superposing multipole fields of different orders. The magnets in a storage ring will usually be constructed so that one particular multipole component is dominant; in the ideal case, the steering magnets (used to control the beam trajectory) will be 'pure' dipoles, and the focusing magnets (used to control the size of the beam) will be 'pure' quadrupoles. In practice, the field in any multipole magnet will contain components from all multipole orders, though one specific order will usually be very much larger than the others. In some cases, the steering magnets will be designed to control both the beam trajectory and the size of the beam: magnets in this case will contain both dipole and quadrupole components of significant strength.

It is also possible to rotate multipole fields about the z-axis. Fields given by the above expression (1.2) with real valued coefficients C_n are known as *normal* multipoles. A rotation of the field through an angle $\pi/2(n + 1)$ gives a *skew* multipole, which can be described by (1.2) with the relevant coefficient C_n having

---

[2] The electromagnetic theory needed to describe the fields in accelerator components can be found in many standard texts, for example [14, 15]. For a discussion more specialised for magnets in accelerators, see [16].

a pure imaginary value. A complex value for C_n (i.e. if C_n has real and imaginary parts) represents a superposition of a normal and skew multipole. A skew dipole will deflect a beam vertically rather than horizontally. A skew quadrupole will allow control over *coupling* in the beam (see section 2.6).

Magnets other than multipole magnets are used in storage rings for particular purposes. For example, strong solenoid fields are often used around the detector in a collider, although the role of the solenoid in this case is connected with the operation of the detector rather than the storage ring itself. *Insertion devices* consisting of sequences of (short) dipole magnets of alternating polarity are used in light sources to enhance the production of synchrotron radiation (see chapter 3).

Accelerator magnets can be electromagnets, or can be constructed from permanent magnetic materials. The advantage of electromagnets is that the field strength can be readily adjusted by controlling the flow of current through the coils of the magnet: although the field strengths in the magnets will be fixed during operation, commissioning and tuning a storage ring generally requires some adjustments to be made to the magnet strengths. The drawback of electromagnets is that high currents (several tens or hundreds of amperes) are usually needed to achieve the specified field strengths, so that providing the power, and cooling the magnets, can be an issue. Superconducting magnets are able to provide significantly higher field strengths than normal-conducting magnets; however, the additional cost and complexity associated with the cryogenics system needed to operate superconducting magnets means that normal-conducting magnets are usually preferred, where possible. Although magnets with adjustable field strength can be constructed using permanent magnetic materials, the mechanisms needed to provide the adjustment for such magnets have so far meant that (normal-conducting) electromagnets are usually the preferred option.

1.2.2 Radiofrequency cavities

Radiofrequency (RF) cavities [17] are used in electron storage rings to replace the energy that particles lose through synchrotron radiation. An RF cavity contains an oscillating electromagnetic field, with a dominant electric field component parallel to the trajectory of the beam as it passes through the cavity [18]. The energy gain of a particle of charge q as it passes through the cavity is $q V_0 \cos(\phi)$, where V_0 is the peak voltage across the cavity, and the particle arrives at a phase ϕ of the field oscillation. Although the energy lost in a single turn of the ring is usually only a small fraction of the energy of a particle, several cavities with peak voltages of the order of a megavolt or more are usually needed to maintain a beam in an electron storage ring operating with beam energy of a few GeV. The maximum voltage that can be achieved in a cavity is limited by the point at which electrons are stripped from the inner surface of the cavity, leading to field breakdown.

The oscillation frequency of the field in the RF cavities must be matched to the revolution frequency of particles in the ring: this is the basic principle behind operation of a synchrotron. A small change in the RF frequency will lead to a change in the beam energy, which will in turn change the revolution frequency: the dependence of the revolution frequency on the particle energy is characterised by the *phase slip factor* of the ring (see section 2.7) and allows stable operation of a storage ring even if the RF frequency is not set perfectly.

The phase of the RF field oscillation at which the voltage across the cavity exactly matches the energy lost by a particle to synchrotron radiation is known as the *synchronous phase*. If particles arrive at a phase slightly ahead of, or behind, the synchronous phase, then their motion around the ring can remain stable, through the mechanism of *phase stability* (see section 2.8). However, there are limits on the maximum distance from the synchronous phase for which stable motion can be maintained: particles arriving too far from the synchronous phase will be unable to adjust their energy while remaining within the storage ring, and will be lost from the beam. As a result, the beam in an electron storage ring will consist of bunches of particles, separated by gaps with length corresponding to the RF oscillation period. Typically, a bunch in a storage ring will be of order 10 ps (a few millimetres) in length, and the gaps between bunches will be of order 2 ns (about two-thirds of a metre), corresponding to an RF frequency of 500 MHz.

The shape of the field within each RF cavity must be carefully controlled to minimise any adverse impacts on beam behaviour. The oscillating electric field will induce a magnetic field in the cavity; in a simple cavity with a geometry that is approximately cylindrical, the electric field will be parallel to the axis of the cylinder, and the magnetic field lines will form circular loops centred on the axis. The magnetic field can deflect particles passing through the cavity, and since the strength of the magnetic field increases with distance from the axis, this can lead to focusing effects.

In addition to the 'fundamental' mode in which a longitudinal electric field accelerates particles passing through the cavity, the fields in an RF cavity can occur in *higher-order modes* in which the fields oscillate at higher frequencies, and form different patterns within the cavity. The fundamental mode is driven by electro-magnetic fields generated by the RF power supply (such as a klystron, or a solid-state amplifier) and fed into the cavity through a waveguide and RF coupler. However, higher-order modes can be driven by the electromagnetic fields around the particles in the beam as they travel through the cavity. In some circumstances, the higher-order modes can reach amplitudes large enough that particle trajectories are deflected by an amount large enough for the beam to become unstable (see chapter 5). The cavity modes (the fundamental mode, as well as the higher-order modes) occur at discrete frequencies and with field patterns determined by the boundary conditions set by the shape of the cavity. If the frequencies of any of the higher-order modes coincide with frequencies present in the beam current spectrum, then resonances can occur in which the higher-order modes are driven to large amplitudes. An important step in the design of a storage ring is the optimisation of the design of the RF cavities, to avoid as far as possible any overlap between the cavity mode spectrum and the beam current spectrum. Nevertheless, the parameter regimes specified for modern electron storage rings can be extremely challenging, and feedback systems (as outlined below, in section 1.2.3) are often needed to maintain beam stability.

To minimise the dissipation of RF power in the walls of the cavity, RF cavities must be made from materials with a good electrical conductivity. The material must

Figure 1.4. Radiofrequency cavity (copper chamber) in the storage ring of the Advanced Light Source at the Lawrence Berkeley National Laboratory. As it emits synchrotron light, the electron beam loses energy, which must be replaced if the beam is to continue circulating in the storage ring. The lost energy is put back into the beam by RF cavities, such as the copper structure in the centre of this photograph. Image courtesy of the Advanced Light Source, Lawrence Berkeley National Laboratory. Copyright 2010 The Regents of the University of California, through the Lawrence Berkeley National Laboratory.

also have appropriate mechanical and thermal properties to allow easy fabrication and stable operation. For cavities designed to operate at room temperature, copper is a common choice: see figure 1.4. However, at high field strengths, the amount of power dissipated in the walls of the cavity through induced electrical currents can be considerable, and it is usually necessary to provide cooling, by means of water flowing through pipes or channels fitted around the cavity. Some electron storage rings use superconducting RF cavities, which have the advantage of achieving relatively low power dissipation because of the extremely high conductivity of superconducting materials (although the DC resistivity of a superconductor is zero, oscillating currents are associated with some energy dissipation). The rate of decay of an oscillating electromagnetic field in a superconducting cavity can be slower by several orders of magnitude than the rate of decay in a comparable normal conducting cavity. The drawback with superconducting cavities is that it is necessary to operate below the critical temperature for the material from which the cavity is made. Although high-temperature superconducting materials do exist, their mechanical properties make them unsuitable for the fabrication of RF cavities. Most superconducting RF cavities are made from niobium[3], which has a critical temperature of 9.2 K. However, the operating temperature needs to be significantly

---

[3] Niobium has the highest critical temperature of any elemental superconductor.

lower than this (typically, around 4.5 K) so that the cavity remains superconducting in the presence of the strong magnetic fields that are inherent in the function of the cavity. The need for a cryogenic system, which brings additional cost and complexity to the RF system, often outweighs the benefits of superconducting technology: the choice between copper and niobium cavities is not a straightforward one.

1.2.3 Feedback systems

Storage rings usually contain many feedback systems with a wide variety of functions and operating over different timescales and parameter regimes. In general, the purpose of a feedback system is to maintain the stability of a measured parameter by making adjustments to components controlling that parameter. As an example, consider the beam trajectory (i.e. the orbit) in a storage ring: it is important that this is tightly controlled in a light source so that the beams of synchrotron radiation are directed accurately through the radiation beamlines to the experimental stations. In a collider, control over the orbit is needed to maintain good luminosity. The orbit is measured using a set of beam position monitors (BPMs) distributed around the ring, and can be adjusted using small dipole magnets that provide horizontal or vertical corrections to the beam trajectory. Over different timescales, small changes in orbit can occur from a number of causes, including (for example) electrical noise on the magnet power supplies, mechanical vibrations from pumps or other equipment, and temperature variations causing magnet supports to expand or contract. The first step to take to maintain orbit stability is to minimise the environmental effects affecting the orbit: magnet power supplies need to be of high quality (low noise), pumps should be mounted on supports that provide mechanical isolation from any vibration, magnet supports should be designed to minimise sensitivity to temperature variations, and the temperature in the tunnel housing the accelerator should be kept stable (in practice, often to a fraction of one degree Celsius). However, some residual beam motion is inevitable, so a feedback system is used to monitor the beam position around the ring, and to determine and apply the appropriate corrections to the strengths of the steering magnets to maintain a pre-defined orbit.

Calculating the appropriate correction to be applied by a feedback system from a set of measurements is not always a straightforward process. In the case of an orbit feedback system, for example, the way that the orbit responds to given changes in strength of the steering magnets depends on the strengths of all the other magnets in the storage ring. If the strengths of these magnets change slightly, then the orbit may respond in an unexpected way to changes in the steering magnets. If the feedback system is not carefully designed, then it may respond by attempting further corrections of steadily increasing amplitude, with the orbit becoming worse at each step until the beam is eventually lost from the storage ring. The same consequences can result if the signals from the BPMs are noisy, so that the correction applied by the feedback system is based on inaccurate data. Feedback systems are often needed on systems that are inherently unstable, and can be very sensitive to small environmental changes; an effective and reliable feedback system must be carefully designed, based on a good understanding of the system on which it will work. A number of standard algorithms have been developed, which can be applied in different situations; see, for example, [19].

Beyond achieving the required stability, a further challenge in many cases is the timescale on which a feedback system may need to operate. In the case of an orbit feedback system, orbit changes in response to ground motion or temperature variations may be on the timescale of minutes, hours or days: such timescales do not pose significant problems for modern feedback systems. However, electrical noise may cause oscillations in beam position with frequencies of the order of hundreds of hertz, or several kilohertz. Collecting data from a large number (maybe dozens) of BPMs, then calculating and applying a correction within a few milliseconds can be very challenging. In a storage ring, beam instabilities driven by wake fields (electromagnetic fields within the beam pipe generated by the beam itself) can occur on the timescale of a few microseconds: it is possible, using specialised fast feedback systems, to suppress such instabilities, allowing storage rings to operate in parameter regimes that would otherwise be inaccessible. Fast (bunch-by-bunch) feedback systems for suppressing beam instabilities are discussed further in section 5.6.

Other feedback systems in a storage ring may serve to maintain beam optics parameters (such as the betatron tunes), the beam energy and intensity; or they may operate within technical subsystems, for example to maintain the stability of the current from a magnet power supply, or to improve the stability of the frequency and amplitude of the fields in an RF cavity.

1.2.4 Vacuum systems

Even at very low pressure, gas in the beam pipe in an electron storage ring can have a number of effects that limit machine performance [20]. Particles in the beam may collide with gas molecules, leading to a loss of beam current. Gas molecules may become ionised by collisions, and the resulting positive ions can be 'trapped' in the negative electrical potential around a beam of electrons; interactions between the electrons and the ions can then cause the beam to become unstable. In positron storage rings, electrons from the ionisation of gas molecules can collide with the walls of the beam pipe, releasing additional electrons. Under some circumstances, the density of electrons (the *electron cloud*) can build up to the point where the beam becomes unstable. In colliders, scattering of particles in the beam from gas molecules in the vicinity of the interaction region can lead to backgrounds in the detector.

Gas pressures inside the vacuum chamber of a particle accelerator are commonly measured in torr (Torr): standard atmospheric pressure is 760 Torr (or 1.01×10^5 Pa in SI units). The pressure in the vacuum chamber of an electron storage ring is typically specified to be of order 1 nTorr, or roughly 10^{-12} times atmospheric pressure. At this pressure, the beam lifetime (the time taken for the beam current to fall by a given factor) is likely to be dominated not by gas scattering, but by other effects, in particular Touschek scattering (section 5.1). In practice, the pressure in a storage ring can vary considerably depending on the location around the ring and the operational conditions. When an electron beam is present, synchrotron radiation falling on the walls of the vacuum chamber can release gas molecules through photodesorption, leading to a significant increase in the local pressure. For this reason, the vacuum chamber is designed to minimise the amount of synchrotron radiation falling on the inner wall of the chamber. In a light source, the intense radiation beams from insertion devices are generally directed to radiation beamlines; but there can still be considerable amounts of radiation, for example from dipoles, remaining within the beam pipe. The fact that radiation can be reflected multiple times from the walls of the beam pipe can make it difficult to predict exactly where photodesorption will occur. A common technique to remove radiation from the main chamber is to shape the vacuum chamber so that it has a slot running along one side that opens into an 'antechamber'. The slot allows radiation to pass out of the main chamber, but since it has a lower conductance than the main chamber for the flow of gas, it is possible to maintain a lower pressure in the main chamber than in the antechamber.

The specified pressure within the vacuum chamber of an electron storage ring is usually only achieved after a long period (maybe some weeks or months) of pumping and conditioning. Different types of pumps are needed to cover the range of pressures and species in the residual gas. For example, turbo-molecular pumps can operate from atmospheric pressure to below 0.01 nTorr, and can pump any kind of gas molecule. However, the pumping is localised so that the required pressure may only be achieved in the near vicinity of the pump; this limitation can be overcome by installing large numbers of pumps, but their high cost can make this prohibitive. Titanium sublimation pumps can achieve high pumping speed at relatively low cost, and can be implemented to provide 'distributed' pumping, i.e. operating to pump long sections of the beam pipe. However, titanium sublimation pumps cannot operate above pressures of around 10^{-4} Torr, and will not pump molecules of inert gases such as argon and methane. Similarly, different types of instrumentation (pressure gauges) are needed to ensure accurate readings of pressure in different regimes and environments.

Materials used in the construction of an accelerator vacuum chamber must have certain properties. For example, the surface of any material will absorb gas molecules when exposed to gas at high (atmospheric) pressure, then release the gas molecules when the pressure is reduced (in a vacuum vessel) in a process known as *outgassing*. The outgassing rates from materials used in an accelerator beam pipe should be low enough to allow the specified pressure to be maintained without excessive numbers of pumps. The chamber walls should have good electrical conductivity to minimise the electromagnetic fields generated by particles in the beam as they travel through the chamber (in particular, to minimise resistive-wall wake fields, see section 5.3). The material should have good mechanical properties to allow the different parts of the chamber to be fabricated easily. In positron storage rings, it may be necessary to choose a material with a low probability of releasing electrons, either from photoemission or from the incidence of primary electrons or ions, to avoid the build-up of an electron cloud (section 6.2.3). Finally, the chamber should be able to withstand heating by a couple of hundred degrees, which may be required to condition the chamber or to activate certain types of coating (in particular, non-evaporable getter, or NEG coatings) used to provide additional pumping. Aluminium is a common choice of material for the vacuum chambers of electron storage rings. Some of the properties of aluminium are not ideal; for example, it can lose mechanical strength when heated to the temperatures need to activate NEG coatings. Nevertheless, aluminium can meet many of the requirements for the vacuum chamber in an electron storage ring, at a reasonable cost.

The geometry of the vacuum chamber is also important for a number of different reasons. First, the aperture should be large enough to accommodate the beam, allowing for particles performing transverse oscillations of large amplitude, or with significant energy errors. Steering errors and processes occurring during injection can also lead to large deviations of the beam trajectory from the ideal orbit. From point of view of the vacuum, a large aperture helps to achieve a good gas conductance through the chamber, which can improve the pumping rate achieved with a fixed number of pumps. Second, the chamber should be designed to allow as much synchrotron radiation as possible to leave the main chamber without striking the walls; the impact of synchrotron radiation on the walls of a vacuum chamber can lead to photodesorption, or to local heating. The operation of some beam diagnostic devices can be affected by synchrotron radiation, and may need to be shielded in some way. In a light source, extraction ports are necessary to allow the radiation to exit the ring towards experimental areas, but ports may also be necessary to extract the synchrotron radiation from colliders. Third, the chamber needs to allow some flexibility to allow for changes in position or length in different sections, in response to the motion of components attached to the chamber or to changes in temperature. Some components fixed to the chamber (for example, beam position monitors) also need to be isolated as far as possible from vibrations or other mechanical motion that may be transmitted along the chamber. Flexibility and a certain amount of mechanical isolation can be provided by 'bellows' located at appropriate points around the ring. Finally, the geometry of the inside of the chamber needs to be as smooth as possible, avoiding gaps or sudden changes between different apertures or cross-sections. This is because abrupt transitions can trap electromagnetic fields generated by particles in the beam, leading either to localised heating as the energy of the fields is dissipated in the walls of the chamber, or to wake fields that act back on the beam potentially causing beam instabilities.

The requirements on the chamber geometry are often in conflict; for example, it can be difficult to provide ports for vacuum pumps or for extracting synchrotron radiation while maintaining the 'smoothness' necessary to minimise wake fields. Bellows also tend to be associated with cavities that can lead to strong wake fields. The solution is often to provide some 'RF shielding' using strips of metal arranged to present a smooth, unbroken screen to electromagnetic fields oscillating at microwave (and lower) frequencies, while allowing the relevant components (vacuum pumps, bellows, etc) to perform their appropriate function. Optimising the design of a vacuum chamber can be a complex, iterative task involving modelling the vacuum, mechanical, and electromagnetic properties, and often requiring some compromises to be made to achieve a practical solution. For further information on a range of aspects of vacuum systems for accelerators, see [21].

1.2.5 Diagnostics

Beam diagnostics [22] are critical for the commissioning and effective operation of a storage ring. The behaviour of the beam in a storage ring is sensitive to a large number of environmental factors, and can be affected in particular by field and alignment errors on the magnets, frequency and amplitude errors in the RF cavities, and timing errors on the injection system. A variety of collective effects (some of which are discussed further in chapter 5) can impact the beam parameters and behaviour in ways that can be difficult to predict, and that depend on the beam intensity, the magnetic lattice and the design of the vacuum chamber. No matter how carefully a storage ring is designed and assembled, injecting and storing a beam and achieving the specified beam quality and stability depends on detailed, accurate, and complete information on the properties and behaviour of the beam. Any storage ring must include a wide range of diagnostics devices to provide information on the charge in individual bunches, the position of the beam at different points around the storage ring, and the transverse beam size and bunch length. The design of the diagnostics system needs to take into account the most critical points at which particular measurements need to be made, the parameter range over which the diagnostics should provide accurate information, the rate at which measurements need to be taken, and the impact that the diagnostics devices themselves will have on the beam. Installing additional diagnostics after construction and commissioning can be difficult, so it is important to include the diagnostics system as an integral part of the design process for a storage ring.

Many diagnostic devices work on the principle of directly detecting the electromagnetic fields around a bunch of charged particles; but the detection can be done in many different ways. For example, to measure the total amount of charge in a bunch, a metal block can be placed in the path of the bunch. When the charge hits the block, the electric potential of the block changes by $V = Q/C$ where Q is the charge on the bunch and C is the capacitance of the block. Unfortunately, this technique destroys the bunch in the process of making the measurement, so is usually not suitable for a storage ring. It can, however, be a useful method in a single-pass accelerator, such as a linac, or in the transfer lines used for injecting a beam into a storage ring. In practice, the block needs to be carefully shaped so that as many electrons as possible are trapped within the block, rather than simply being scattered around it: this generally leads to a *Faraday cup* design, in which a bunch passes through an aperture into a cavity within the block. Furthermore, the voltage signal can be very small and last for a very short time, so the electronics used to detect the signal need to be very sensitive, with low noise, and high bandwidth (i.e. able to process signals that change rapidly with time). Within a storage ring, charge measurements are more routinely made using either a *DC current transformer* (DCCT) to measure the voltage induced in a coil of wire by the magnetic field around the beam, or a *wall-current monitor* placed across an insulating gap in the vacuum chamber to detect the electric currents induced in the chamber by the beam. These devices are non-destructive, in the sense that they allow the beam to continue to circulate during and following the measurement.

After measurement of the bunch charge, the trajectory of the beam is usually the next critical measurement needed during commissioning and tuning of a storage ring. In general, the trajectory is measured using a series of *beam position monitors* (BPMs), each of which consists of a set of electrodes that protrude into the vacuum chamber but are insulated from it. A simple BPM pickup may consist of one electrode protruding into the top of the chamber, and another protruding from the bottom. As a bunch passes between the electrodes, it will induce a voltage on each electrode, with the size of the voltage depending on the proximity of the beam to the electrode: the closer the beam to the electrode, the larger the voltage. Thus, the difference between the voltage signals is related to the vertical position of the beam in the chamber. The sum of the voltages induced on the electrodes can give an indication of the total charge in the bunch, although specialised devices (such as a DCCT or a wall-current monitor) can usually provide more accurate and reliable charge measurements. Of course, the voltages induced on the electrodes also depend on the horizontal position of the bunch within the vacuum chamber; a common arrangement in a storage ring is therefore to use four electrodes to measure the horizontal and vertical position of the beam at the same time. The shape of the electrodes must be optimised to maximise the sensitivity of the BPM, and to minimise wake fields (see section 5.3): in electron storage rings 'button' electrodes are commonly used, in which the electrode terminates in a flat plate that is flush with the surrounding surface of the vacuum chamber. Electrodes consisting of 'strip lines' can also be used; the optimum geometry depends on the parameter regime. Very precise beam position measurements (with resolution less than 1 μm) can be made from the oscillating fields induced when a bunch passes through a cavity. However, the wake fields resulting from cavity BPMs usually preclude their use in storage rings.

Interpreting the signals from a beam position monitor is not completely straightforward. Typically, the BPMs in a modern electron storage ring are required to provide measurements of the position of the beam centroid with a resolution of order 10 μm (which may be comparable to the transverse size of the beam). The variation of the voltage on each electrode with beam position will be nonlinear, and will depend on the shape of the surrounding vacuum chamber. The voltage will also depend on the bunch charge. The signal returned by the electronics used to process the voltages on the electrodes will be subject to some level of noise, and will also depend on temperature. Determining the beam position from the signals returned by a BPM requires careful calibration, and will be subject to random and systematic errors. Random errors (from noise in the electronics) can be reduced by making repeated measurements, although it may be difficult to distinguish between real changes in beam position and random fluctuations in the BPM signals. Systematic errors (for example, from effects associated with the local geometry of the vacuum chamber, or temperature variations within the electronics) can be more difficult to characterise. Readings from a BPM are often best regarded as providing relative rather than absolute measurements of beam position.

Measurements of the transverse beam size can be obtained by placing a screen in the path of the beam. The screen may be made from a material that fluoresces when hit by a bunch of charged particles, such as yttrium aluminium garnet (YAG); or, it may be a simple metal foil from which *transition radiation* may be observed from a charged bunch crossing the boundary from the vacuum to the interior of the metal. In either case, the light produced can be observed with a camera. A screen using transition radiation has the advantage that the radiation pulse length is equal to the length of the electron bunch producing the radiation (typically of order several picoseconds), whereas the light from a YAG screen can persist for a considerable time (maybe hundreds of nanoseconds) after the bunch has passed. Thus, a screen using transition radiation can, with a suitable fast camera, provide information on the size of an individual bunch, whereas a YAG screen may, depending on the bunch spacing, provide a signal representing an integration over many bunches. On the other hand, a YAG screen is usually more sensitive than a screen using transition radiation, and may therefore be more appropriate at relatively low beam currents and beam energies. Both techniques are, of course, destructive, and therefore not suitable for a storage ring in regular operation. Instead, measurements of beam size may be made from observation of the synchrotron radiation produced from charged particles moving through magnetic fields. Observing the radiation is straightforward, but again, obtaining the desired information (in this case, of the size of the beam producing the radiation) requires a detailed understanding of the physical principles and practical implementation of the various components of the system, and careful analysis of the signals obtained.

Measurements of bunch length are also important in optimising the performance of an electron storage ring, and can (like measurements of the transverse beam size) be performed using synchrotron radiation. A common instrument used for bunch length measurements is a *streak camera*. A streak camera consists of a plate that emits electrons when struck by a pulse of radiation. The electrons are accelerated (by a static electric field) towards a detector, but between the emitter and the detector pass through a transverse electric field that is oscillating at high frequency. The variation of the transverse field means that the electrons receive a transverse deflection with a magnitude that depends on the time at which they pass through the field. Early electrons then arrive at the detector at a different transverse position from late electrons. The transverse spread of the electrons on the detector provides a measurement of the length of the electron bunch in the storage ring. The importance of synchrotron radiation for electron beam diagnostics is such that electron storage rings often have synchrotron radiation beamlines dedicated to diagnostics for the beam in the storage ring.

A wide range of diagnostics, beyond those mentioned here, have been developed to provide more complete and detailed information on beam properties in storage rings (and other types of accelerator). New diagnostic techniques are still being developed, either to widen the scope of information available, or to provide better precision or accuracy in various parameter regimes than can be achieved using existing techniques, or to minimise the effects of the measurement on the beam. In some cases, valuable information can be obtained by using a range of diagnostics while manipulating the beam in various ways. For example, the position of a beam in a quadrupole magnet can be found by changing the strength of the magnet while observing the beam orbit using the BPMs; this method, sometimes called *beam-based alignment*, can be used to steer the beam through the centres of the quadrupole magnets, which is often an important step in tuning an electron storage ring for optimum performance. Any tuning procedure requires not only suitable diagnostics, but also a good understanding of those diagnostics: this will involve not just an appreciation of the physical principles on which the measurement is based, but also an awareness of the appropriate parameter regimes, the limitations of the diagnostics in terms of accuracy and the sources and sizes of random and systematic errors, the ways in which any measurements from the diagnostics may be verified, and the potential impact on the beam. Often, the best approach is to regard the beam and the diagnostics as a single physical system, and to view the signals from the diagnostics in that context rather than taking them at face value.

1.2.6 Control systems

The purpose of the control system [23] in an accelerator is to allow the operators (through a convenient interface) to set the parameters for the various components and subsystems, to read signals from beam diagnostics and other components, and to display and record these signals as required. A modern storage ring can include thousands of separate components, and the inevitable complexity of a system on such a scale means that careful consideration needs to be given to the design, implementation, and maintenance of the control system. Typically, the architecture of a control system consists of three layers. In the bottom layer are computers that communicate directly with a small number of components (such as magnet power supplies, beam position monitors, vacuum pumps etc). The top layer consists of computers that display information for the operators, and allow the operators to send commands to the various components and subsystems. The middle layer consists of a set of powerful data servers that channel information between the top and bottom layers, and record data from different sources. The required data transfer rates may be very large, for example in the case of high-resolution camera images collected at high frame rates; the data files that need to be recorded may similarly be of a considerable size. The computer hardware and software need to be carefully selected to match the performance requirements.

In practice, a modular approach is often the most practical, with each of the different layers structured to reflect different sets of components or subsystems in the accelerator. Different software systems may be used in different parts of the control system; there are then potential issues, for example regarding compatibility of data formats. However, tools are generally available that allow different platforms to communicate with each other efficiently.

1.2.7 Injection systems

The beam in an electron storage ring follows a closed orbit, in which each bunch of particles returns to the same position after each turn of the ring. An injection system must control the trajectory of a bunch arriving at the storage ring, so that the new bunch follows the closed orbit once it is in the ring [24]. The simplest system conceptually is 'single-turn' injection, in which a dipole magnet is located at the point where the trajectory of an incoming bunch crosses the closed orbit. Under normal circumstances, this dipole magnet is turned off, so that it does not deflect bunches that are already on the closed orbit. However, during injection the dipole is turned on, so that an incoming bunch, initially travelling at some angle to the closed orbit, is deflected so that it follows the closed orbit on leaving the dipole. It is then necessary to switch off the dipole before another bunch already on the closed orbit arrives, otherwise this bunch will be deflected out of the storage ring. This scheme requires a dipole magnet that can be turned on and off very quickly: a magnet of this type is often called a *kicker magnet*. Even with a fast kicker magnet, single-turn injection usually means leaving a gap in the train of bunches long enough for the field in the kicker to be turned on and off. A gap in the bunch train in an electron storage ring may be needed in any case, to avoid ion trapping (see section 5.2). However, another problem with single-turn injection is that the source must be capable of delivering in a single pulse a train of bunches each with the full charge specified for each bunch in the storage ring. This can place severe demands on the particle source.

An alternative injection scheme involves filling the storage ring over several turns, merging additional charge with the charge already in the storage ring. This can be accomplished using a *septum magnet*. A septum magnet consists of a dipole magnet with two apertures separated by a narrow plate, or 'blade'. By passing an electric current through the blade, it is possible to arrange for the dipole field to appear only in one of the apertures. The aperture on the other side of the blade, in which there is zero field, is the aperture through which the beam passes when on the closed orbit. The injection process follows a number of steps. First, a set of dipoles is used to move the closed orbit so that it comes very close to the septum blade. Then, a bunch from the source is directed through a transfer line towards the storage ring, where it arrives at the septum on the side of the blade where there is a strong dipole field. The field deflects the incoming bunch so that it is travelling parallel to the closed orbit, but at some displacement from it. Since the injected bunch is not on the closed orbit, it will oscillate around the closed orbit as it travels around the ring (see section 2.2). The magnets in the storage ring are set so that once the bunch arrives back at the septum after completing a turn of the storage ring, it is at a phase of its orbit oscillation such that it passes through the septum through the field-free aperture. The final step in the process is then to turn off the dipole magnets that were used to distort the closed orbit, so that (usually after several turns) the beam follows its original closed orbit. Meanwhile, the oscillations of the newly injected bunch around the closed orbit will steadily decrease in amplitude as a result of processes such as synchrotron radiation damping (see section 3.2.2). Thus, after several turns, the charge in the ring has increased and the beam is once again on the original closed orbit. This process can be repeated as often as necessary to reach the full beam current needed in the storage ring. Although dipoles are still needed that can be turned on and off quickly, the timescale is now on the order of several revolution periods, rather than a fraction of a revolution period. The demands on the particle source are also greatly relaxed, compared to single-turn injection. A drawback with multi-turn injection is that there is a risk of significant beam losses on the septum, or other components: the septum blade in particular can easily be damaged if the heat load from particles lost from the injected or stored beams is too high. The injection system needs to be carefully designed and the entire storage ring needs to be finely tuned to minimise beam losses during injection.

There are a number of variations on the scheme outlined above for multi-turn injection, which may have advantages in some cases, usually by reducing injection losses. For example, it is possible to take advantage of the variation in particle trajectory with the energy of the particle (an effect known as *dispersion*—see section 2.5) by injecting bunches with some difference in energy: a 'dispersion bump' can then be used in addition to (or even instead of) an 'orbit bump' in the region of the septum.

Particles in the beam in an electron storage ring can collide with residual gas particles in the vacuum chamber or with other particles in the beam (see section 5.1): these collisions or scattering events can result in the loss of particles from the beam, leading to a decay in the beam current. If no additional particles are injected into the ring to replace the particles lost by scattering, the beam current can fall by half over the course of a few hours; the exact timescale can vary quite widely, depending on the beam parameters and the pressure of the residual gas in the vacuum chamber. Traditionally, electron storage rings have been operated in such a way as to refill the ring a couple of times each day. This reduces the impact on users from variations in beam orbit and beam size associated with the injection process; however, it has the disadvantage that there are large variations in the brightness of the radiation beams (in a synchrotron light source) or the luminosity (in a collider) over time. Also, the heat load on different components in the accelerator can change substantially in response to changes in current, making it difficult to maintain temperature stability. To address these problems, light sources have developed ways of operating with 'top-up' injection, in which small amounts of charge are injected at short, regular intervals of perhaps no more than a few minutes. If the effects on the beam from the injection process can be kept small enough, then top-up injection provides better overall beam stability for users than the traditional mode of refilling at intervals of several hours.

1.2.8 Personnel protection

Accelerators present a number of hazards, including components or substances at very high or very low temperature, high voltages, toxic chemicals, confined spaces (where there may in some cases be a risk of oxygen depletion from the accidental release of liquified gases), laser beams, and radiation. Systems need to be in place to minimise the risks from all hazards. The first step should be to eliminate each hazard if at all possible, for example by designing systems to avoid the possibility of components reaching temperatures high enough to present a risk of burns. Inevitably, however, there will be instances where hazards cannot be avoided, and precautions must then be taken to reduce the risks. Possible measures may include: enclosing or shielding the hazard; limiting access to only those personnel with a need to work in the area affected, and who have been properly trained and authorised; ensuring that clear warning notices are displayed; and providing monitoring systems and alarms The use of personal protective equipment should usually be the last line of defence.

Although all hazards need to be properly considered and addressed, systems associated with protection from radiation are often the most visible in an accelerator facility [25]. Radiation may be generated from any high-energy particles that are incident on matter; the type of radiation and its properties will depend on the type of particle involved, the particle energy, and the materials struck by the particles. Synchrotron radiation from relativistic particles undergoing acceleration (see chapter 3) is also a hazard. In electron storage rings, the highest levels of radiation are produced while the accelerator is operating. The ring will be enclosed either in an underground tunnel, or within a 'surface tunnel' constructed from large concrete blocks. Access to the tunnel will be through a limited number of entrances, usually with a labyrinth so there is no direct line of sight to the accelerator from outside the tunnel. Each entrance will have an interlocked gate or door, so that the accelerator is automatically and immediately switched off if anyone tries to enter the tunnel during operation. Before the accelerator is turned on, the tunnel will be searched according to a well-defined procedure to ensure that there is no-one in the tunnel before the entrance interlocks are activated. In the case of an electron storage ring, after the accelerator is turned off the radiation from any materials that have been activated will normally decay quite rapidly; nevertheless, measures will be taken to ensure that radiation in the area around the accelerator is below specified limits before allowing general access.

Personnel protection systems need to be completely reliable. For that reason, systems (including alarms and interlocks) are designed with significant redundancy in different components, and so that any failure will lead to a safe condition; for example, an interlock on the entrance to the accelerator tunnel should actively report when the entrance gate is closed, so that if the interlock fails the system interprets the lack of a signal as meaning that the gate is open. Although the personnel protection system may communicate with the accelerator control system, the need for high reliability often means that critical safety systems are controlled directly by dedicated hardware, and are not under computer (software) control.

1.3 Some examples of electron storage rings

Although electron storage rings in third-generation synchrotron light sources have a number of common features, they can also vary quite widely in their structure and parameters. It is difficult to identify any one ring as 'typical'; however, the parameters for some representative examples are given in table 1.1. Even in the cases given, the parameters can be changed depending on the needs of the users. For example, some light source users need radiation in short pulses (tens of picoseconds) with relatively long gaps (hundreds of nanoseconds) in between: a time structure of this kind can be achieved by injecting charge into the ring in just one or two bunches. The synchrotron radiation produced by the beam in this case will have low average

Table 1.1. Parameters of some storage rings for third-generation synchrotron light sources. It is important to note that most light sources have significant flexibility in their operating parameters, so as to be able to vary the characteristics of the synchrotron radiation that they produce for users. Parameters marked with an asterisk (*) in the table can be varied, in some cases over a very wide range: the values given should be taken as being indicative.

	ALS	SPring-8	AS	DLS	MAX IV
Start of operation	1993	1999	2007	2007	2017
Circumference (m)	197	1436	216	562	528
Lattice type	TBA	DBA	DBA	DBA	7BA
Number of arc cells	12	44	14	24	20
Beam energy* (GeV)	1.9	8.0	3.0	3.0	3.0
Beam current* (mA)	500	100	200	300	500
Natural emittance* (nm)	2.0	2.4	7.1	3.2	0.3
RF frequency (MHz)	500	509	500	500	100
RF voltage* (MV)	1.1	16	3.2	2.5	1.5
Bunch length* (ps)	21	13	22	20	33
Beam lifetime* (h)	8	200	16	10	10

intensity, because the charge that can be contained in a single bunch is limited by collective effects. Other users may be less sensitive to the time structure of the radiation, but need a high average intensity: for these users, as much charge as possible will be injected into the ring, usually in several hundred bunches separated by gaps of a few nanoseconds. In addition to changes in filling pattern, there can be variations in the beam optics that can affect the beam size, which is determined (in part) by the emittance of the beam[4]. Storage rings used in light sources also often have upgrade programmes intended to extend the life of the facility by taking advantage of developments in technology that allow improved performance. Individual upgrade projects may range in scale from the installation of a small number of new components, to the replacement of essentially the entire storage ring.

The Advanced Light Source (ALS [26], see figure 1.3) in Berkeley, California, USA is one of the earliest third-generation light sources, i.e. one of the first light sources to be developed with a low-emittance lattice for the production of high-brightness synchrotron radiation. The low emittance is achieved by the use of a triple-bend achromat (TBA) structure for the lattice (see section 3.3): each arc cell includes three dipole magnets. Although other light sources have been built using TBA arc cells, the double-bend achromat (DBA, with just two dipoles in each arc cell) has been a more popular choice. Recently, however, there has been interest in lattice styles using a larger number of dipoles in each arc cell, to achieve very low emittances: an example is MAX IV in Lund, Sweden [27, 28].

---

[4] Emittance is discussed in more detail in section 2.3; but briefly, emittance is the product of the beam size and divergence (the rate of change of the beam size). Emittance is a more useful measure of beam quality than beam size, since the emittance of a beam is constant as the beam moves around a storage ring, whereas the beam size varies depending on the local focusing provided by quadrupole magnets.

SPring-8 (Super Photon Ring—8 GeV [29]) in Hyogo Prefecture, Japan, is one of five light sources around the world operating with electron beam energies above 5 GeV. Using a higher energy for the electron beam makes it easier to generate high-brightness synchrotron radiation at shorter wavelengths (i.e. in the hard x-ray region of the electromagnetic spectrum), which is needed for some applications. There are also some benefits of higher energy in terms of the accelerator physics: for example, the beam becomes less sensitive to certain collective effects, and tends to have a longer lifetime. The drawback of a higher beam energy is that to achieve a low emittance, which is necessary for generating high-brightness synchrotron radiation, a large circumference, accommodating a large number of arc cells, is needed (see section 3.3). A number of the subsystems, in particular the RF system, also need to be larger in rings operating at higher energy. The result is that both the construction cost and the running costs of the facility increase with increasing energy. Many third-generation light sources are designed to operate with a beam energy around 3 GeV, which provides a good compromise between the capability to produce high-brightness, short-wavelength synchrotron radiation and the cost of the facility.

The Australian Synchrotron (AS [30]) in Melbourne, Australia, and the Diamond Light Source (DLS [31], figure 1.5) in Didcot, UK, are examples of synchrotron light sources constructed in the first decade of the twenty-first century. These, and many other facilities constructed around the same period, are based on double-bend achromat lattices, with 3 GeV beam energy. There can still be significant differences between such machines, however. DLS, for example, has more than twice the

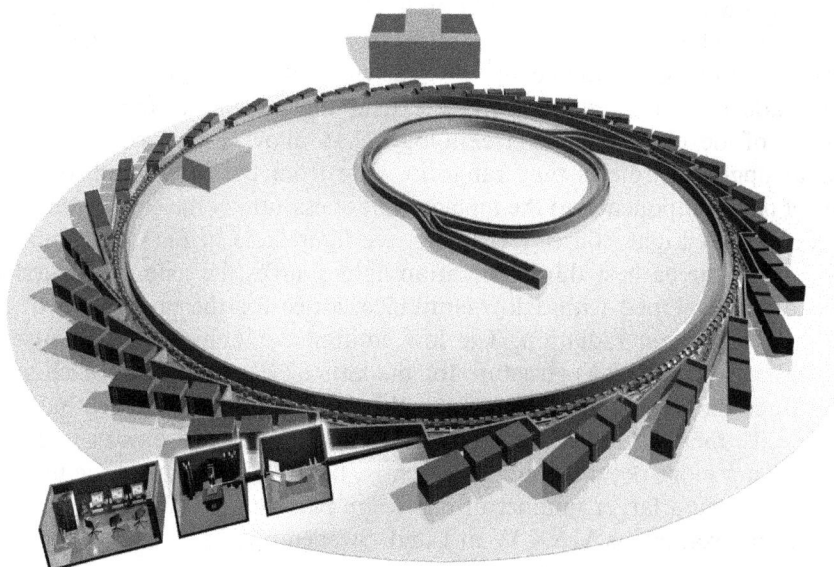

Figure 1.5. Schematic of the DLS. The electron source and booster synchrotron (for raising the particle energy before injection into the storage ring) are located within the main storage ring. Synchrotron radiation beamlines, constructed tangential to the main storage ring, carry synchrotron radiation from dipoles and insertion devices to the experimental areas. Image courtesy of Diamond Light Source.

Table 1.2. Parameters of some storage rings for electron–positron colliders. As with light sources, storage rings for colliders usually have significant flexibility in their operating parameters. Parameters marked with an asterisk (*) in the table can be varied, in some cases over a very wide range: the values given should be taken as being indicative.

	CESR	LEP	DAΦNE	SuperKEKB	
				LER (e^+)	HER(e^-)
Start of operation	1979	1989	1999	2018	2018
Circumference (m)	768	26 659	98	3016	3016
Beam energy* (GeV)	5	104	0.51	4	7
Beam current* (mA)	365	5.2	5000	3600	2600
Natural emittance* (nm)	300	21	1000	3.2	2.4
RF frequency (MHz)	500	352	368	509	509
RF voltage* (MV)	5	3500	0.254	8.4	6.7
Bunch length* (mm)	20	10	30	6	5
Beam lifetime* (h)	5	5	2	0.2	0.2

circumference of AS. This not only allows the ring to accommodate a larger number of arc cells, enabling a lower emittance (and higher brightness for the synchrotron radiation beams), but also provides for a much larger number of radiation beamlines, thus accommodating a larger number of user experiments.

Storage rings in colliders are in many aspects more varied than the storage rings used in light sources. Table 1.2 shows some of the parameters for a number of electron–positron colliders. The scientific goals for a collider tend to be more specific than for light sources, which serve a more diverse user community. Broadly speaking, colliders aim to discover new phenomena either by achieving higher centre-of-mass (collision) energies than previous colliders, or by achieving higher luminosities. The collision energy determines the processes that can occur during a collision between two particles; for a process to occur, the collision energy will normally have to exceed some threshold. The luminosity determines the rate at which the various processes occur in a collider: the higher the luminosity, the more events of a given type will be observed, allowing more precise measurements of the parameters associated with that event (for example, the mass of a particle produced in a collision).

Large Electron–Positron Collider (LEP), at CERN, Geneva, Switzerland [32] operated from 1989 to 2000, reaching the highest collision energy ever achieved in an e^+e^- collider. The final beam energy of 104 GeV was only possible because of the large circumference of the machine. The energy lost from the beam through synchrotron radiation has to be replaced by the RF cavities; the larger the circumference, the lower the synchrotron radiation power, and the lower the cost of the RF system. Hadron colliders can achieve much higher collision energies because the synchrotron radiation losses depend on the mass of the radiating particle: the heavier the particle, the smaller the radiation power. The drawback with hadron colliders is that the collisions are far more complex than in lepton colliders, because of the internal structure of hadrons, and the ability of hadrons to interact via the strong nuclear force. It is unlikely that any future electron–positron colliders based on storage rings will exceed the centre-of-mass energy achieved in LEP. There have been extensive studies of linear colliders that avoid the problem of synchrotron radiation by accelerating and colliding electrons and positrons in a machine in which the beam essentially follows a straight path; however, there are significant technical difficulties that would need to be overcome in such a machine.

Other storage ring colliders, including the Cornell Electron Storage Ring (CESR, at Cornell University, Ithaca, New York, USA [33, 34]), DAΦNE (the Double-Annulus Phi factory for Nice Experiments, at LNF, Frascati, Italy [35]) and SuperKEKB (Super KEK B-factory, at KEK in Tsukuba, Ibaraki Prefecture, Japan [36]) aim for maximum luminosity, rather than the highest possible energy. Each of these machines is optimised for a particular process, or a small number of processes, to be studied in detail. For example, SuperKEKB is intended to study the physics of B mesons: the centre-of-mass beam energy is set at the peak of the cross-section (i.e. the probability) for production of $\Upsilon(4S)$ mesons, which decay to B mesons. Of particular interest is the asymmetry between the B meson and its antiparticle. The asymmetry in the electron and positron beam energies in SuperKEKB (and in the earlier colliders, PEP-II and KEKB) allows precise measurements of the lifetimes of the particle–antiparticle pairs of B mesons produced in the collisions.

The very wide variation between parameters that can be seen from the examples in table 1.2 results from the fact that the different colliders are optimised for different physics goals. Each different parameter regime tends to involve different challenges for the design, construction, and operation of the storage ring.

References

[1] Sessler A and Wilson E 2007 *Engines of Discovery: A Century of Particle Accelerators* (Singapore: World Scientific)

[2] Rubensson J-E 2016 *Synchrotron Radiation: An Everyday Application of Special Relativity* (San Rafael, California, USA: Morgan & Claypool Publishers) (Bristol, UK: IOP Concise Physics, IOP Publishing)

[3] Rowe E M and Mills F E 1973 Tantalus I: a dedicated storage ring synchrotron radiation source *Part. Accel.* **4** 211–27

[4] Winick H and Doniach S (ed) 1980 *Synchrotron Radiation Research* (New York, NY, USA: Plenum)

[5] Clarke J A 2004 *The Science and Technology of Undulators and Wigglers* (Oxford, UK: Oxford University Press)

[6] Freund H P and Antonsen T M Jr 1996 *Principles of Free-electron Lasers* II edn (London, UK: Chapman and Hall)

[7] Saldin E L, Schneidmiller E A and Yurkov M V 2008 *The Physics of Free Electron Lasers* (Berlin, Germany: Springer)

[8] Szarmes E B 2014 *Classical Theory of Free-electron Lasers* (San Rafael, California, USA: Morgan & Claypool Publishers) (Bristol, UK: IOP Concise Physics, IOP Publishing)

[9] Kim K-J, Huang Z and Lindberg R 2017 *Synchrotron Radiation and Free-electron Lasers* (Cambridge, UK: Cambridge University Press)

[10] Willmott P 2011 *An Introduction to Synchrotron Radiation: Techniques and Applications* (Chichester, UK: Wiley)

[11] Mobillo S, Boscherini F and Meneghini C (ed) 2015 *Synchrotron Radiation: Basics, Methods and Applications* (Heidelberg, Germany: Springer)

[12] Margaritondo G 2002 *Elements of Synchrotron Light for Biology, Chemistry and Medical Research* (Oxford, UK: Oxford University Press)

[13] Winick H (ed) 1994 *Synchrotron Radiation Sources: A Primer* (Singapore: World Scientific)

[14] Griffiths D J 2017 *Introduction to Electrodynamics* IV edn (Cambridge, UK: Cambridge University Press)

[15] Jackson J D 1998 *Classical Electrodynamics* III edn (New York: Wiley)

[16] Wolski A 2014 October Maxwell's equations for magnets *Technical Report* arXiv:1103.0713v2

[17] Vretenar M 2011 Radio frequency for particle accelerators—evolution and anatomy of a technology *Proc. CERN Accelerator School 2010: RF for accelerators (Ebeltoft, Denmark, 8–17 June 2010)* ed B Roger CERN–2011–007 (Geneva, Switzerland: CERN), pp 1–14

[18] Wolski A 2011 Theory of electromagnetic fields *Proc. CERN Accelerator School 2010: RF for accelerators (Ebeltoft, Denmark, 8–17 June 2010)* ed B Roger CERN–2011–007 (Geneva, Switzerland: CERN), pp 15–66

[19] Taylor C J, Young P and Chotai A 2013 *True Digital Control: Statistical Modelling and Non-minimal State Space Design* (New York: Wiley)

[20] Mistry N B and Li Y 2013 Requirements for vacuum systems *Handbook of Accelerator Physics and Engineering* II edn ed A Wu Chao, K Hubert Mess, M Tigner and F Zimmermann (Singapore: World Scientific), pp 431–32

[21] Brandt D (ed) 2007 June *Proc. CERN Accelerator School Course on Vacuum in Accelerators (Platja d'Aro, Spain, 16-24 May 2006). Technical Report CERN-2007-003* (Geneva, Switzerland: CERN)

[22] Brandt D (ed) 2009 August *Proc. CERN Accelerator School Course on Beam diagnostics (Dourdan, France, 28 May-6 June 2008). Technical Report CERN-2009-005* (Geneva, Switzerland: CERN)

[23] Rehlich K 2013 Controls and timing *Handbook of Accelerator Physics and Engineering* II edn ed A Wu Chao, K Hubert Mess, A Wu Chao, M Tigner and F Zimmermann (Singapore: World Scientific), 760–3

[24] Rees G H 2013 Ring injection and extraction *Handbook of Accelerator Physics and Engineering* II edn ed A Wu Chao, K Hubert Mess, M Tigner and F Zimmermann (Singapore: World Scientific)

[25] Roesler S and Silari M 2013 Radiation protection principles *Handbook of Accelerator Physics and Engineering* II edn ed A Wu Chao, K Hubert Mess, M Tigner and F Zimmermann (Singapore: World Scientific), pp 767–9

[26] Advanced Light Source https://als.lbl.gov [Online; accessed 29 October 2017]

[27] MAX IV https://www.maxiv.lu.se [Online; accessed 21 March 2018]

[28] Leemann S C, Sjöström M and Andersson Å 2017 May First optics and beam dynamics studies on the MAX IV 3 GeV storage ring *Proc. Eighth International Particle Accelerator Conference (Copenhagen, Denmark)* pp 2756–59

[29] SPring–8 http://www.spring8.or.jp/en/ [Online; accessed 29 October 2017]

[30] Australian Synchrotron http://www.synchrotron.org.au [Online; accessed 20 March 2018]

[31] Diamond Light Source http://www.diamond.ac.uk [Online; accessed 20 March 2018].

[32] Aßmann R, Lamont M and Myers S 2002 A brief history of the LEP collider *Nucl. Phys. B (Proc. Suppl.)* **109B** 17–31

[33] Berkelman K 2004 *A Personal History of CESR and CLEO* (Singapore: World Scientific)

[34] CESR https://www.classe.cornell.edu/Research/CESR/WebHome.html [Online; accessed 21 March 2018]

[35] DAΦNE http://www.lnf.infn.it/acceleratori/ [Online; accessed 21 March 2018]

[36] SuperKEKB http://www-superkekb.kek.jp/documents.html [Online; accessed 20 March 2018]

Chapter 2

Linear optics

There are many similarities between the way that a charged particle moves through an accelerator beamline and the way in which a ray of light moves through a system of prisms and lenses. Just as a prism can be used to change the direction of a light ray, a dipole magnet can be used to steer the trajectory of a charged particle. A set of light rays can be focused (or defocused) by a lens; the trajectories of a set of charged particles can be focused (or defocused) by a quadrupole magnet. However, there are also some significant differences between the way in which particle trajectories can be controlled using magnets, and the way in which light rays can be controlled using prisms and lenses. For example, it is possible in the case of light rays to use a lens to focus the rays simultaneously in the horizontal (transverse) and vertical directions; but as we shall see in section 2.1.3, a quadrupole magnet that focuses a beam in one direction (either horizontally or vertically) will necessarily defocus in the other direction. Despite these differences, the rules describing the motion of charged particles through beamlines consisting of dipole and quadrupole magnets, together with a few other types of component, form a framework known as 'charged particle beam optics'. If we make some simplifications so that the equations describing the motion of particles through a beamline are linear equations, then we are discussing linear charged particle beam optics, or simply 'linear optics'. In this chapter, we outline the principles and key results from linear optics. Although we do not give rigorous derivations, the material is standard, and is presented in detail in a number of texts on accelerator beam dynamics, see for example [1–5].

2.1 Co-ordinate system and transfer matrices

A comprehensive description of the motion of particles through an accelerator can be very complicated if all significant effects are taken into account. However, for many purposes, it is sufficient to consider how individual particles move through

doi:10.1088/978-1-6817-4989-1ch2   2-1

drift spaces (where there are no electric or magnetic fields), dipole magnets, quadrupole magnets, and radiofrequency (RF) cavities. This is the case, for example, when first laying out the design of a storage ring, or at an early stage of commissioning. To a good approximation, the motion of particles through these components can be described by linear equations. It is then possible and convenient to write the equations relating the co-ordinates of a particle at the exit of a component to the co-ordinates at the entrance in terms of a matrix, known as a *transfer matrix*. The elements of a transfer matrix are determined by the type of component and the relevant parameters, e.g. the length and field strength of a dipole magnet.

In this section, we shall give formulae for the transfer matrices for drift spaces, dipole magnets, quadrupole magnets and RF cavities. However, we first need to define the co-ordinate system and the variables that we will use to describe the motion of particles moving through these elements. It should be noted that different definitions are in use amongst the accelerator community, and there can be apparent differences in the description of the physics, depending on which set is used. The choice of co-ordinates for any given situation is to some extent a question of taste or convenience; and ultimately, of course, all valid choices should give the same results in terms of the observable behaviour of particles in the accelerator.

Let us first consider a co-ordinate system in a quadrupole magnet: the magnetic field in this case is shown in figure 2.1. The strength of the field is proportional to the

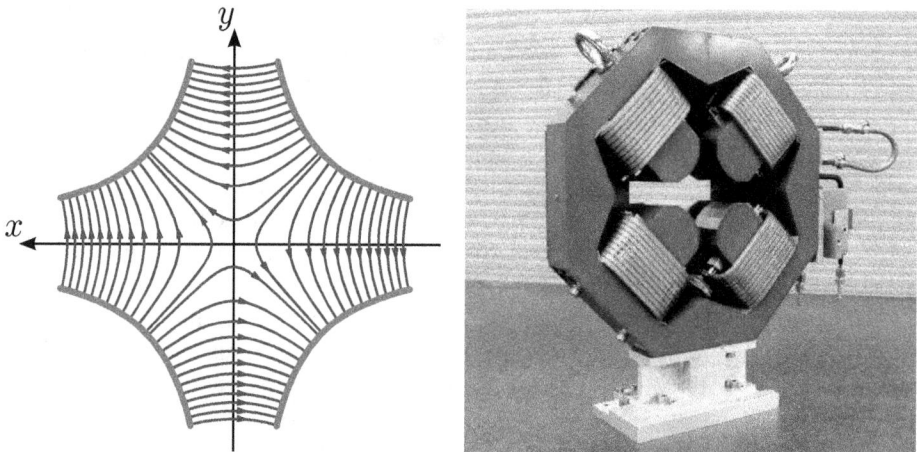

Figure 2.1. Magnetic field in a quadrupole magnet. The diagram on the left shows the field lines (in blue) in a cross-section of a quadrupole magnet. The field is zero along the magnetic axis, which is a line passing perpendicularly into the page at $x = y = 0$. The arrows show the direction of the field along the field lines, which are drawn only in the space between the poles of the magnet; the pole faces are shown in blue. The strength of the field increases linearly with distance from the magnetic axis, and (within the main body of the magnet) is independent of distance along the axis. The picture on the right shows an example of an electromagnetic quadrupole. The steel yoke and poles are coloured red; the copper windings carry electric current. Accelerator magnets often require large currents (tens or hundreds of amps) to achieve the specified field strengths, and the coils then take the form of pipes carrying cooling water to prevent the magnet from overheating. Reproduced with permission from Science and Technology Facilities Council.

distance from the *magnetic axis*, which is a line passing through the centre of the magnet. The field strength at any point along the magnetic axis is zero. In most cases, the ideal trajectory of a particle through a quadrupole magnet is along the axis. In practice, however, particles will always enter the magnet with some transverse displacement from the axis: the problem then is to determine the position of each particle with respect to the axis as it exits the magnet. At any point in the magnet we can imagine a plane perpendicular to the axis, and we can specify any point in this plane using Cartesian co-ordinates, with the origin defined by the point where the magnetic axis passes through the plane. In what follows, we shall use the convention that the x co-ordinate of a particle is the position along the horizontal axis in this co-ordinate system, and the y co-ordinate is the position along the vertical axis. The distance along the magnetic axis at which we construct the plane we denote by the variable s.

We are now in a position to describe the trajectory taken by a particle as it moves through a quadrupole magnet: to do so, we have to give the x and y co-ordinates of a particle as functions of s. In this case of course, the co-ordinates x and y and the variable s together define a three-dimensional Cartesian co-ordinate system; however, this will not be the case for all types of magnet. In particular, in a dipole magnet there is no line along which the field strength is zero: the field (by definition) is the same strength everywhere within the magnet. This means that in a dipole, it is more convenient to define an arc of a circle as the equivalent of the axis in a quadrupole magnet: the distance along the arc we again denote by s, and although the line described by different values of s is now curved we can again at any point along this line construct a plane transverse to the line, with Cartesian co-ordinates x and y. The variables x, y and s now define the position of a particle in a curvilinear co-ordinate system. The reason for choosing an arc of a circle as the origin of the transverse co-ordinates in a dipole is that the ideal trajectory for a particle moving through a dipole is along a circular arc: the line taken by a particle with the 'correct' momentum will be given by $x(s) = y(s) = 0$. This greatly simplifies the description of the motion of particles moving through the magnet, even if they do not follow the ideal trajectory.

So far, we have considered a co-ordinate system that can be used within each individual component in an accelerator. But we can generalise the discussion to a sequence of components forming an accelerator beamline. To specify the position of a particle at any point along the beamline, we first define the *reference trajectory* which can, in principle, be any line in space. We would be wise, however, to define the geometry of this line so that it follows the magnetic axis in each quadrupole magnet, and an arc of suitable radius in each dipole magnet. In drift spaces, where there is no magnetic field, we can choose the reference trajectory to follow the line joining the reference trajectory at the exit of one magnet with the reference trajectory at the entrance of the next magnet. Note that the reference trajectory is not necessarily a line followed by a real physical particle in an accelerator: it is simply a line defined for our own convenience, to specify the positions of particles in the accelerator. Usually, however, we do define the reference trajectory so that a particle with the appropriate momentum will follow this trajectory along the entire beam-line. Although this simplifies the description of the motion of particles in the accelerator, identifying the reference trajectory with the physical trajectory of a particle can lead to some confusion, and it is helpful to maintain a clear concept of the reference trajectory as something separate from the actual physical trajectory of a particular particle.

2.1.1 Drift spaces

We can now return to the use of transfer matrices to describe (and calculate) the trajectories of particles along an accelerator beamline. First, consider a particle in a drift space. At the start of the drift space (defined by $s = s_0$, for some position s_0 along the reference trajectory), the particle has horizontal transverse[1] co-ordinate $x(s_0)$, and horizontal transverse momentum $P_x(s_0)$. At the end of the drift space (defined by $s = s_1$) the particle has horizontal transverse co-ordinate $x(s_1)$ and horizontal transverse momentum $P_x(s_1)$. Since there are no electric or magnetic fields in the drift space, and we ignore gravity, there are no forces on the particle, so there is no change in its momentum. Therefore, $P_x(s_1) = P_x(s_0)$. However, the horizontal co-ordinate of the particle will (in general) change, depending on the angle at which the particle is travelling with respect to the reference trajectory. If the particle has total momentum P_T and is travelling *almost* parallel to the reference trajectory, then the angle between the particle trajectory and the reference trajectory is given to a good approximation by $\theta \approx P_x(s_0)/P_T$. The change in the horizontal co-ordinate between the entrance and exit of the drift space will then be given by $\Delta x = L \tan(\theta) \approx L P_x(s_0)/P_T$, where $L = s_1 - s_0$ is the length of the drift space. To summarise, the horizontal co-ordinate and momentum at the end of the drift space are related to the horizontal co-ordinate and momentum at the entrance by:

$$x(s_1) \approx x(s_0) + L\frac{P_x(s_0)}{P_T}, \tag{2.1}$$

$$P_x(s_1) = P_x(s_0). \tag{2.2}$$

For convenience, we define a *reference momentum* P_0, which is chosen so that each particle in the beamline will (ideally) have total momentum close to P_0. We can then define the *scaled* horizontal momentum of a particle: $p_x(s) = P_x(s)/P_0$ (note that the scaled momentum is written with a lower-case p). The equations describing the change in horizontal co-ordinate and momentum of a particle over the length of a drift space can then be written in matrix form:

$$\begin{pmatrix} x \\ p_x \end{pmatrix}_{s=s_1} = R_{\text{drift},x} \begin{pmatrix} x \\ p_x \end{pmatrix}_{s=s_0}, \tag{2.3}$$

---

[1] We use the term 'transverse' to refer to directions in the plane perpendicular to the reference trajectory.

where the matrix $R_{\text{drift},x}$ is given by:

$$R_{\text{drift},x} = \begin{pmatrix} 1 & L \\ 0 & 1 \end{pmatrix}. \tag{2.4}$$

$R_{\text{drift},x}$ is the transfer matrix describing the change in the horizontal transverse co-ordinate x and momentum p_x over a drift space of length L.

2.1.2 Dipole magnets

Let us now consider a particle moving through a dipole magnet. If the field is in a vertical direction, and the particle is moving along a reference trajectory that lies in a horizontal plane, then the force on the particle will also be in the horizontal plane, but perpendicular to the reference trajectory[2]. In other words, the force will have only a component F_x parallel to the x-axis; the strength of the force will be $F_x = q\, v\, B$, where q is the electric charge of the particle, v is its velocity, and B is the strength of the magnetic field.

Since the force from the dipole magnetic field will *always* be perpendicular to the direction of motion of the particle, the particle will move along the arc of a circle: the centripetal force is $F = \gamma m v^2/\rho$, where m is the mass of the particle, v is its velocity, and ρ is the radius of curvature of the particle trajectory. The relativistic factor $\gamma = 1/\sqrt{1 - v^2/c^2}$ takes into account the effective increase in mass of a particle as its velocity approaches the speed of light c. If the magnetic field is vertical, then the circular trajectory of the particle will be in the horizontal plane. For a particle in a dipole field, the centripetal force is provided by the magnetic force: we can then say that $qvB = \gamma m v^2/\rho$, and hence:

$$\frac{P_T}{q} = B\rho, \tag{2.5}$$

where $P_T = \gamma m v$ is the total momentum of the particle. The quantity P_T/q, known as the *beam rigidity*, turns out to be very important for describing the motion of particles through magnetic fields in accelerator beamlines. Although it is really a measure of the momentum of a particle, the beam rigidity is often written as $B\rho$ even when there is no particular magnetic field strength B (or trajectory radius ρ) being considered; this serves to emphasise the fact that for a given momentum, the product of the magnetic field and the radius of curvature of the particle trajectory is a constant.

Let us choose, for a given dipole, a reference trajectory with curvature $\rho = P_0/qB$, where B is the dipole field strength, q the charge on the particles, and P_0 is the reference momentum (chosen to be close to the momentum of each particle in the beamline). In that case, a particle with momentum P_0 will follow the reference trajectory through the dipole, if it enters the dipole magnet on the reference trajectory. If a particle enters at some angle to the reference trajectory, i.e. with

---

[2] The direction of the force $\vec{F}$ on a particle with charge q in a magnetic field $\vec{B}$ follows from the vector product in the Lorentz force: $\vec{F} = q\vec{v} \times \vec{B}$, where $\vec{v}$ is the velocity of the particle.

some non-zero horizontal transverse momentum, then the horizontal transverse co-ordinate $x(s)$ of the particle will change as a result, but the horizontal transverse momentum will stay the same. In fact, over a short distance, the transfer matrix for a particle in a dipole magnet will be approximately the same as the transfer matrix for a particle in a drift space of equivalent length. However, over a longer distance, the trajectory of the particle will oscillate around the reference trajectory. This is because, if the magnet is long enough, both the reference trajectory and the physical trajectory of the particle will form complete circles, although the centres of the two circles will be displaced from each other (see figure 2.2). Taking this effect into account, the horizontal co-ordinate $x(s_1)$ and momentum $p_x(s_1)$ at the end of a dipole magnet can be written in terms of the horizontal co-ordinate $x(s_0)$ and momentum $p_x(s_0)$ at the entrance of the magnet as follows [4]:

$$\begin{pmatrix} x \\ p_x \end{pmatrix}_{s=s_1} = R_{\text{dipole},x} \begin{pmatrix} x \\ p_x \end{pmatrix}_{s=s_0} + \vec{m}_{\text{dipole},x}, \tag{2.6}$$

where

$$R_{\text{dipole},x} = \begin{pmatrix} \cos(\omega_x L) & \sin(\omega_x L)/\omega_x \\ -\omega_x \sin(\omega_x L) & \cos(\omega_x L) \end{pmatrix}, \tag{2.7}$$

and

$$\vec{m}_{\text{dipole},x} = \left(\frac{P_T}{qB} - \rho \right) \begin{pmatrix} 1 - \cos(\omega_x L) \\ \omega_x \sin(\omega_x L) \end{pmatrix}. \tag{2.8}$$

In these expressions, L is the length of the dipole magnet, and $\omega_x = \sqrt{qBP_T\rho}$ where P_T/q is the beam rigidity for the particle, B is the dipole field strength, and ρ is the radius of curvature of the reference trajectory. $R_{\text{dipole},x}$ is the transfer matrix for a dipole magnet of length L and field strength B.

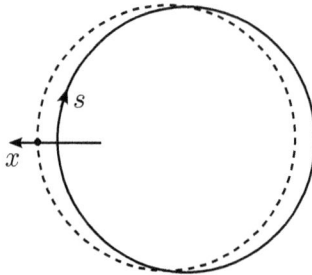

Figure 2.2. Trajectory of a charged particle in a dipole field. The field is perpendicular to the page, and the reference trajectory is shown by a circle (solid line). The distance along the reference trajectory is given by the variable s. At any point $s = s_0$ along the reference trajectory, the distance of the particle from the reference trajectory (in the plane of the reference trajectory) is given by the co-ordinate $x(s_0)$. The particle follows a circular trajectory (dashed line) that oscillates around the reference trajectory as the particle moves through the dipole field, completing one oscillation on each complete turn.

In contrast to the case of a drift space, the equations relating the final co-ordinate and momentum to the initial co-ordinate and momentum have zeroth-order terms (contained in the vector $\vec{m}_{\text{dipole},x}$) as well as first-order terms (contained in the transfer matrix $R_{\text{dipole},x}$). However, if the reference trajectory is chosen so that it is also the physical trajectory for a particle with momentum P_T equal to the reference momentum P_0, then the beam rigidity for the particle is $P_T/q = B\rho$. In that case, we see that both components of $\vec{m}_{\text{dipole},x}$ are zero, and the equation describing particle motion in a dipole (2.6) resembles the equation describing particle motion in a drift space (2.3), though with a different transfer matrix. Since there are no zeroth-order terms in this case, a particle entering the dipole magnet along the reference trajectory will leave the magnet along the reference trajectory. However, in the more general case, where a particle has momentum not exactly equal to the reference momentum, then the elements of $\vec{m}_{\text{dipole},x}$ will be non-zero, and a particle entering the dipole along the reference trajectory will reach the end of the dipole with some horizontal displacement with respect to the reference trajectory. This effect, by which a particle moves away from the reference trajectory in a dipole because of the difference between its momentum and the reference momentum for the beamline, is referred to as *dispersion*, and will be discussed in more detail in section 2.5.

Notice that the transfer matrix for a dipole (2.7) resembles a rotation matrix in a co-ordinate system with axes x and p_x. A co-ordinate system where one axis represents the co-ordinate of a particle and the other axis represents the momentum is usually known as *phase space* (see figure 2.3). The dipole transfer matrix represents the motion of a particle around an ellipse in phase space. In other words, if we represent the co-ordinate and momentum of a particle at the start of a dipole by a point in phase space, the co-ordinate and momentum at the end of the dipole can be found by following the arc of an ellipse around the origin in phase space. If a particle passes directly from one dipole magnet to another, forming a complete circle in real (co-ordinate) space, the motion of the particle can be represented by a complete ellipse in phase space. This behaviour is associated with the fact that as the particle moves through a long dipole, its trajectory oscillates around the reference trajectory. Looking at the transfer matrix as a rotation matrix, the wavelength of these *betatron oscillations* is the distance the particle travels in making one complete rotation in phase space, and is given by $2\pi/\omega_x$. For a particle with momentum equal to the

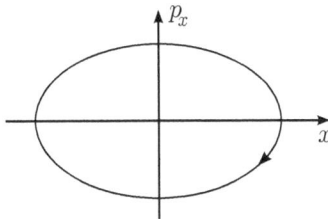

Figure 2.3. Phase space trajectory of a charged particle in a dipole field. The axes in horizontal phase space are the horizontal co-ordinate x and the corresponding momentum p_x. As a particle makes a complete turn in a dipole field (see figure 2.2) its path in phase space forms an ellipse. The direction in which the particle follows the phase space trajectory is shown by the arrow on the ellipse.

reference momentum, $P_T = P_0$, and the betatron frequency is $\omega_x = 1/\rho$; the betatron wavelength is therefore $2\pi\rho$, which is equal to the circumference of the reference trajectory. This is a result that we should expect, given that the particle trajectory is a circle with the same radius as the reference trajectory but a different centre. However, a particle with higher or lower momentum than the reference momentum will follow a trajectory with larger or smaller circumference than the reference trajectory; as a result, the betatron wavelength will change. This is an effect known as *chromaticity*, and will be discussed in more detail in section 4.1.

So far, we have considered only the horizontal motion of particles in drift spaces and dipole magnets, but particles will also move in the vertical direction. In a drift space, the vertical motion behaves in much the same way as the horizontal transverse motion, and can be described by a transfer matrix having the same form as for the horizontal transverse motion. For a particle with vertical co-ordinate $y(s_0)$ at the start of a drift of length L and vertical (scaled) momentum $p_y(s_0)$, the co-ordinate and momentum at the end of the drift are $y(s_1)$ and $p_y(s_1)$, given by:

$$\begin{pmatrix} y \\ p_y \end{pmatrix}_{s=s_1} = \begin{pmatrix} 1 & L \\ 0 & 1 \end{pmatrix}\begin{pmatrix} y \\ p_y \end{pmatrix}_{s=s_0}. \tag{2.9}$$

In a dipole magnet, however, the vertical motion differs from the horizontal. Assuming that the magnetic field is vertical, the force on a particle moving through the field will have zero vertical component. As a result, the transfer matrix for the vertical motion in a dipole is the same as for a drift space—if the particle enters a dipole with non-zero vertical momentum, the vertical distance of the particle from the reference trajectory will steadily increase (until the particle exits the dipole and reaches another magnetic component, or until it hits the wall of the vacuum chamber). This is in contrast to the horizontal motion, where a particle can continue to make stable oscillations around the reference trajectory indefinitely.

2.1.3 Quadrupole magnets

We have seen that in dipole magnets, the horizontal transverse trajectory of a particle can oscillate around the reference trajectory, but the vertical motion will diverge in the same way as in a drift space. Therefore, if we wish to keep particles within the vacuum chamber in an accelerator beamline over a large distance, we need some way to stabilise the vertical motion of the particles. This can be achieved using quadrupole magnets. A quadrupole magnet has a vertical field with a strength that increases in proportion to the horizontal distance from the magnetic axis, and a horizontal field with a strength that increases in proportion to the vertical distance from the magnetic axis (see figure 2.1). Mathematically, the field can be written:

$$B_x = b_1 y, \qquad B_y = b_1 x, \qquad B_s = 0. \tag{2.10}$$

The field has to satisfy Maxwell's equations. This requires that the *same* constant b_1 appears in the expressions for the horizontal and vertical field components.

Consider a particle that enters a quadrupole field with some horizontal displacement with respect to the reference trajectory. As it moves through the magnet, the vertical field will cause a horizontal deflection of the trajectory of the particle. Suppose that the deflection is *towards* the reference trajectory. A particle entering the magnet with a similar size displacement but on the opposite side of the reference trajectory will also be deflected, but because $B_y \propto x$ the field is now in the opposite direction, so the deflection will now also be reversed. Therefore, this second particle will also be deflected *towards* the reference trajectory. Taking into account the fact that the size of the deflection increases in proportion to the distance of a particle from the reference trajectory, we find that for the horizontal motion, a set of particles entering a quadrupole parallel to the reference trajectory but with some range of displacements from it will all be deflected towards a certain point on the reference trajectory—this point is the *focal point* of the magnet. Usually, the focal point is some distance from the end of the magnet, and we can then speak of the length from the end of the magnet to the focal point as the *focal length* of the magnet. In this case, for the horizontal motion, the magnet acts like a converging lens, with the particle trajectories behaving as light rays: see figure 2.4.

In the vertical direction, however, the situation is rather different. If, for horizontal displacements, particles are deflected towards the reference trajectory, then the nature of the field in a quadrupole is such that for vertical displacements, particles are deflected *away* from the reference trajectory—the magnet behaves like a diverging lens rather than a converging lens. The situation can be reversed by changing the sign of the constant b_1 in the equations for the field—the magnet then acts as a converging lens for the vertical motion, but a diverging lens for the horizontal motion.

Unfortunately, because any magnetic field has to satisfy Maxwell's equations, it is not possible to construct a single magnet that will provide horizontal and vertical

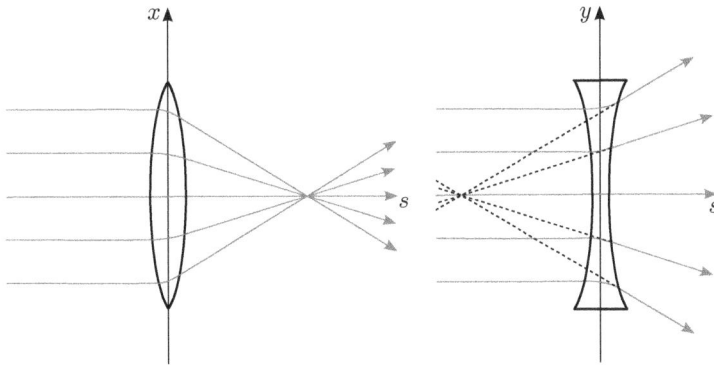

Figure 2.4. Particle trajectories (shown in red) through a horizontally focusing quadrupole magnet. In the horizontal plane (left hand diagram), particles entering parallel to the reference trajectory (along the *s*-axis) but with some displacement from it, will be deflected by the field towards a focal point. In the vertical plane (right-hand diagram) particles initially travelling parallel to the reference trajectory will be deflected by the field so that their trajectories appear to come from a focal point upstream of the magnet. The magnet can be represented by a focusing lens for horizontal particle motion, and a defocusing lens for vertical particle motion. If the magnet is rotated by 90° about the reference trajectory, it becomes a vertically focusing quadrupole, and the effects on the horizontal and vertical particle motion are interchanged.

focusing at the same time. However, as we shall see in section 2.2, stable horizontal and vertical motion in an accelerator can be achieved by using a sequence of quadrupoles that alternate horizontal focusing with vertical focusing.

To derive the transfer matrix for a quadrupole, we proceed as follows. Consider a quadrupole magnet that is short enough so that as a particle moves along the length of the magnet, although its horizontal and vertical momentum will change, any changes in the horizontal co-ordinate x and vertical co-ordinate y are negligible. The change in the horizontal momentum can be found by integrating the horizontal force over the length of the magnet, so that $\Delta p_x = -qvB_y\Delta t/P_0$, where v is the velocity of the particle and $\Delta t = L/v$ is the time taken for the particle to travel the distance L from the start to the end of the magnet. Using $B_y = b_1 x$, we find that $\Delta p_x = -qb_1 xL/P_0$. For convenience, we define the normalised quadrupole gradient k_1:

$$k_1 = \frac{q}{P_0}b_1 = \frac{q}{P_0}\frac{\partial B_y}{\partial x}, \tag{2.11}$$

where P_0/q is the beam rigidity for a particle with the reference momentum P_0 and charge q. Then, the horizontal co-ordinate $x(s_1)$ and momentum $p_x(s_1)$ of a particle at the end of the quadrupole can be written in terms of the co-ordinate $x(s_0)$ and momentum $p_x(s_0)$ at the start as:

$$\begin{pmatrix} x \\ p_x \end{pmatrix}_{s=s_1} = R_{\text{quadrupole},x}\begin{pmatrix} x \\ p_x \end{pmatrix}_{s=s_0}, \tag{2.12}$$

where $R_{\text{quadrupole},x}$ is the horizontal transfer matrix for the quadrupole:

$$R_{\text{quadrupole},x} = \begin{pmatrix} 1 & 0 \\ -k_1 L & 1 \end{pmatrix}. \tag{2.13}$$

If k_1 is positive, then the quadrupole provides horizontal focusing, with focal length $f = 1/k_1 L$. Taking into account the direction of the force on the particle, the vertical transfer matrix for a quadrupole can be written

$$R_{\text{quadrupole},y} = \begin{pmatrix} 1 & 0 \\ k_1 L & 1 \end{pmatrix}. \tag{2.14}$$

To achieve vertical focusing (rather than defocusing) we need a negative value for k_1.

The transfer matrices (2.13) and (2.14) are written assuming that the length of the magnet is short enough that any change in the co-ordinates as a particle moves through the magnet can be neglected. This is known as the *thin lens approximation*. More generally, we need to take the change of co-ordinates into account. A detailed analysis [4] then leads to the horizontal transfer matrix

$$R_{\text{quadrupole},x} = \begin{pmatrix} \cos(\omega L) & \sin(\omega L)/\omega \\ -\omega\sin(\omega L) & \cos(\omega L) \end{pmatrix}, \tag{2.15}$$

where $\omega = \sqrt{k_1}$. For the vertical transfer matrix, we find

$$R_{\text{quadrupole},y} = \begin{pmatrix} \cosh(\omega L) & \sinh(\omega L)/\omega \\ \omega \sinh(\omega L) & \cosh(\omega L) \end{pmatrix}. \qquad (2.16)$$

If k_1 is positive then ω is real, and $R_{\text{quadrupole},x}$ looks like a rotation matrix in phase space. This reflects the fact that the horizontal motion in a sequence of quadrupole magnets all providing horizontal focusing will be stable—the betatron wavelength will be $2\pi/\sqrt{k_1}$. However, the vertical transfer matrix involves hyperbolic sine and cosine functions, rather than regular sine and cosine functions—the vertical motion will be unstable. If k_1 is negative, the situation will be reversed. The value of ω will now be an imaginary number, and the form of the transfer matrices (2.15) and (2.16) will be swapped over, so that the vertical transfer matrix will take the form of a rotation in phase space (stable vertical oscillations) while the horizontal transfer matrix will involve hyperbolic trigonometric functions, reflecting unstable horizontal oscillations.

It is worth noting that the horizontal transfer matrix (2.7) for a dipole is effectively the same as that for a quadrupole (2.15): both a dipole and a quadrupole provide horizontal focusing. However, the betatron wavelength in a quadrupole can be made much shorter than that in a dipole: in a quadrupole, we can achieve a high field gradient b_1 by designing the magnet so that the field changes rapidly over a short distance, even if the peak field strength is relatively low. In a dipole, a short betatron wavelength requires a high field, which can be difficult to achieve. The focusing provided by dipole magnets is therefore usually known as *weak focusing*, whereas that from quadrupole magnets is known as *strong focusing*.

2.1.4 Radiofrequency cavities

In drift spaces, dipole magnets and quadrupole magnets it is often sufficient to describe particle motion in terms of the transverse (x and y) co-ordinates and the corresponding momenta (p_x and p_y) at the start and end of each component. In particular, if the magnetic fields in these components do not change with time, which is the usual case in a storage ring, then to track the trajectory of a particle through a beamline consisting of these components we do not need to know the time at which the particle arrives at each component. However, accelerators also often include components in which the fields do change with time: electron storage rings, for example, must include radio-frequency (RF) cavities to replace the energy lost by particles through synchrotron radiation (see section 1.2.2). An RF cavity is designed to provide a longitudinal electric field that changes the energy of a particle passing through the cavity [6]. To a first approximation, an RF cavity can be represented by a length of beamline across which there is a voltage V that varies sinusoidally in time, so that $V = V_0 \sin(2\pi f t)$. The frequency f is often in the microwave or radiofrequency region of the electromagnetic spectrum, typically several hundred megahertz or a few gigahertz. The effect of the electric field in the cavity on a particle depends on the time at which the particle arrives at the cavity. We therefore need to extend our transfer matrices to include information on the time at which a particle arrives at a given position s along the reference trajectory, as well as the transverse co-ordinates of the particle at that position.

In a storage ring, while the change in the energy of a particle in an RF cavity depends on the time that it arrives at the cavity, the converse is also true: the time at which the particle arrives at a cavity depends on its energy. This is because the trajectory of a particle in a dipole magnet will have a greater length for higher energy particles than for lower energy particles; however, for ultra-relativistic particles, the particle velocity is close to the speed of light and has only a weak dependence on the energy of the particle. Therefore, as long as the particle energy is high enough for the particles to be ultra-relativistic, an increase in the energy of a particle in a storage ring will mean that it takes longer to complete one turn of the storage ring. The time at which a particle arrives at a given position along the reference trajectory and the energy of the particle are therefore connected to each other in much the same way as the transverse co-ordinates x and y are connected to the transverse momenta p_x and p_y.

In principle, we can complete the set of variables needed to describe the motion of a particle in an accelerator by giving, in addition to the co-ordinates x, y and momenta p_x, p_y, the time at which the particle arrives at a given location along the reference trajectory and its energy. However, it is more convenient to work with variables that take small values as the given particle moves along a beamline—the time of arrival at a particular location will (eventually) take a very large value. We therefore specify the time of arrival relative to a notional *reference particle*, which is defined to be a particle moving along the reference trajectory at a speed corresponding to the reference momentum P_0. As we discussed earlier, the reference trajectory is not necessarily a possible physical trajectory for any particle. This implies that a reference particle, which by definition travels exactly along the reference trajectory, cannot (in many cases) physically exist; however, this does not matter for our purposes of describing the motion of real physical particles. The reference particle can be thought of simply as a point travelling along the reference trajectory at a speed $v_0 = P_0/\gamma_0 m$, where P_0 is the reference momentum and $\gamma_0 = 1/\sqrt{1 - v_0^2/c^2}$. We then define the *longitudinal co-ordinate* z of a particle in the accelerator by:

$$z(s) = ct_0(s) - ct(s),\tag{2.17}$$

where $t_0(s)$ is the time at which the reference particle arrives at a location s along the reference trajectory, and $t(s)$ is the time at which the particle with co-ordinate $z(s)$ arrives at the same point. Although we refer to $z(s)$ as a 'co-ordinate' and it has units of distance, it should be remembered that it actually represents a *time*. We similarly define a variable $\delta(s)$, called the *energy deviation*, to specify the energy of a particle relative to the energy E_0 of the reference particle:

$$\delta(s) = \frac{E(s) - E_0}{\beta_0 E_0},\tag{2.18}$$

where the factor $\beta_0 = v_0/c$ is included by convention[3].

---

[3] By defining the energy deviation in this way, the variables $z(s)$ and $\delta(s)$ form a canonical conjugate pair in the Hamiltonian formalism of classical mechanics [7]. This ensures that the variables have certain useful properties in a more formal treatment of particle dynamics [4].

The variables $z(s)$ and $\delta(s)$ describe the *longitudinal* motion of a particle moving along an accelerator beamline. We can write the change in these variables for a particle moving through an RF cavity as follows:

$$\begin{pmatrix} z \\ \delta \end{pmatrix}_{s=s_1} = R_{\mathrm{RF},z} \begin{pmatrix} z \\ \delta \end{pmatrix}_{s=s_0} + \vec{m}_{\mathrm{RF},z}. \qquad (2.19)$$

The general expressions for the elements of the transfer matrix $R_{\mathrm{RF},z}$ and the vector $\vec{m}_{\mathrm{RF},z}$ are rather complicated; but in the special case of an ultra-relativistic particle (large γ) and a cavity that has a length equal to half the RF wavelength, these elements are given to a good approximation [4] by

$$R_{\mathrm{RF},z} = \begin{pmatrix} 1 & 0 \\ -\dfrac{qV_0}{P_0 c} k \cos(\phi_0) & 1 \end{pmatrix}, \qquad (2.20)$$

and

$$\vec{m}_{\mathrm{RF},z} = \begin{pmatrix} 0 \\ \dfrac{qV_0}{P_0 c} \sin(\phi_0) \end{pmatrix}. \qquad (2.21)$$

In these expressions, $k = 2\pi f / c$ where f is the RF frequency, and ϕ_0 is the phase of the RF voltage at which the reference particle (with $z = 0$) arrives at the cavity.

2.1.5 Transfer matrices in three degrees of freedom

Up to now, we have assumed that the horizontal, vertical and longitudinal motion of a particle moving along an accelerator beamline can be treated independently. This is true for an 'ideal' beamline consisting only of drifts and quadrupoles, but as we have already observed, in a dipole magnet the horizontal motion depends on the energy deviation. To take this into account, instead of writing the changes in the transverse and longitudinal variables in terms of three separate 2×2 transfer matrices, we write a single 6×6 transfer matrix to give the change in all six dynamical variables. In general, we can write

$$\begin{pmatrix} x \\ p_x \\ y \\ p_y \\ z \\ \delta \end{pmatrix}_{s=s_1} = R \begin{pmatrix} x \\ p_x \\ y \\ p_y \\ z \\ \delta \end{pmatrix}_{s=s_0} + \vec{m}, \qquad (2.22)$$

where R is a 6×6 transfer matrix, and $\vec{m}$ is a six-component vector. For a dipole of length L [4]

$$R_{\text{dipole}} = \begin{pmatrix} \cos(\omega_x L) & \dfrac{\sin(\omega_x L)}{\omega_x} & 0 & 0 & 0 & \dfrac{1 - \cos(\omega_x L)}{\beta_0 \rho \omega_x^2} \\[2ex] -\omega_x \sin(\omega_x L) & \cos(\omega_x L) & 0 & 0 & 0 & \dfrac{\sin(\omega_x L)}{\beta_0 \rho \omega_x} \\[2ex] 0 & 0 & 1 & L & 0 & 0 \\[2ex] 0 & 0 & 0 & 1 & 0 & 0 \\[2ex] -\dfrac{\sin(\omega_x L)}{\beta_0 \rho \omega_x} & -\dfrac{1 - \cos(\omega_x L)}{\beta_0 \rho \omega_x^2} & 0 & 0 & 1 & \dfrac{L}{\beta_0^2 \gamma_0^2} - \dfrac{\omega_x L - \sin(\omega_x L)}{\beta_0^2 \rho^2 \omega_x} \\[2ex] 0 & 0 & 0 & 0 & 0 & 1 \end{pmatrix}, \quad (2.23)$$

and

$$\vec{m}_{\text{dipole}} = \begin{pmatrix} \left(\dfrac{1}{\rho} - k_0 \right) \dfrac{1 - \cos(\omega_x L)}{\omega_x^2} \\[3ex] \left(\dfrac{1}{\rho} - k_0 \right) \dfrac{\sin(\omega_x L)}{\omega_x} \\[3ex] 0 \\[1ex] 0 \\[1ex] 0 \end{pmatrix}. \quad (2.24)$$

Here, $\omega_x = \sqrt{k_0/\rho}$, where ρ is the radius of curvature of the reference trajectory, and $k_0 = qB_0/P_0$ for dipole field B_0, particle charge q, and reference momentum P_0. If the curvature of the reference trajectory is correctly matched to the dipole field, so that $k_0 = 1/\rho$, then all the zeroth-order terms (i.e. the elements of $\vec{m}$) vanish. This means that a particle with momentum equal to the reference momentum will remain on the reference trajectory if it enters the dipole along the reference trajectory. The non-zero terms in the top right-hand corner of the transfer matrix R_{dipole} relate the change in the transverse variables $x(s)$ and $p_x(s)$ to the energy deviation $\delta(s)$—these elements are associated with the dispersion. We also see that there are non-zero terms in the bottom left-hand corner of the transfer matrix, which relate the change in the variable $z(s)$ (the time at which a particle arrives at a given point along the reference trajectory, relative to the reference particle) to $x(s)$ and $p_x(s)$. These elements are again related to the dispersion, and result from the fact that the length of the trajectory taken by a particle through a dipole depends on its horizontal offset with respect to the reference trajectory—a particle with larger $x(s)$ follows an arc with a larger radius of curvature than a particle with a smaller $x(s)$. Finally, notice that the change in $z(s)$ depends on the energy deviation $\delta(s)$ (the R_{56} element of the transfer matrix is non-zero). The first term in R_{56}, $L/\beta_0^2 \gamma_0^2$, takes into account the fact that, even for ultra-relativistic particles, there is some change in the velocity of a particle (and therefore in the time taken for the particle to travel through the dipole) with a change in the energy of the particle. For particles moving close to the speed of light, however, γ_0 is large, so this term will be small.

The full 6×6 transfer matrix for a drift can be constructed by assembling the 2×2 transfer matrices for the horizontal, vertical and longitudinal variables:

$$R_{\text{drift}} = \begin{pmatrix} 1 & L & 0 & 0 & 0 & 0 \\ 0 & 1 & 0 & 0 & 0 & 0 \\ 0 & 0 & 1 & L & 0 & 0 \\ 0 & 0 & 0 & 1 & 0 & 0 \\ 0 & 0 & 0 & 0 & 1 & \dfrac{L}{\beta_0^2 \gamma_0^2} \\ 0 & 0 & 0 & 0 & 0 & 1 \end{pmatrix}. \tag{2.25}$$

The zeroth-order terms in the transfer map for a drift space vanish (i.e. all elements of $\vec{m}_{\text{drift}}$ are zero). For a quadrupole, the transfer matrix is:

$$R_{\text{quadrupole}} =$$
$$\begin{pmatrix} \cos(\omega L) & \sin(\omega L)/\omega & 0 & 0 & 0 & 0 \\ -\omega \sin(\omega L) & \cos(\omega L) & 0 & 0 & 0 & 0 \\ 0 & 0 & \cosh(\omega L) & \sinh(\omega L)/\omega & 0 & 0 \\ 0 & 0 & \omega \sinh(\omega L) & \cosh(\omega L) & 0 & 0 \\ 0 & 0 & 0 & 0 & 1 & \dfrac{L}{\beta_0^2 \gamma_0^2} \\ 0 & 0 & 0 & 0 & 0 & 1 \end{pmatrix}, \tag{2.26}$$

where L is the length of the quadrupole, and $\omega = \sqrt{k_1}$ with

$$k_1 = \frac{q}{P_0} \frac{\partial B_y}{\partial x}. \tag{2.27}$$

The zeroth-order terms in the transfer map vanish.

In the case of an RF cavity, it turns out that in addition to the change in the energy deviation, there are changes in the transverse variables that resemble horizontal and vertical focusing from a quadrupole [8, 9]: this is because the oscillating electric field generates a magnetic field (to satisfy Maxwell's equations) that deflects particles in the radial direction. The full 6×6 transfer matrix for an RF cavity is then

$$R_{RF} =$$

$$
\begin{pmatrix}
\cos(\omega L) & \sin(\omega L)/\omega & 0 & 0 & 0 & 0 \\
-\omega \sin(\omega L) & \cos(\omega L) & 0 & 0 & 0 & 0 \\
0 & 0 & \cos(\omega L) & \sin(\omega L)/\omega & 0 & 0 \\
0 & 0 & -\omega \sin(\omega L) & \cos(\omega L) & 0 & 0 \\
0 & 0 & 0 & 0 & 1 & \dfrac{L}{\beta_0^2 \gamma_0^2} \\
0 & 0 & 0 & 0 & -\dfrac{qV_0}{P_0 c} k \cos(\phi_0) & 1
\end{pmatrix},
\qquad (2.28)
$$

where L is the length of the cavity, and

$$
\omega = k \sqrt{\frac{qV_0}{2\pi P_0 c} \cos(\phi_0)}. \qquad (2.29)
$$

As before, V_0 is the amplitude of the voltage in the RF cavity, ϕ_0 is the phase at which the reference particle arrives at the cavity, and $k = 2\pi f/c$ where f is the RF frequency. The zeroth-order terms in the transfer map for an RF cavity all vanish.

As we shall see in section 2.2, by multiplying the transfer matrices for individual components in a beamline, we can construct the transfer matrix for any section of the beamline. This provides a powerful tool for understanding and analysing the motion of particles through beamlines that may be of considerable length and contain large numbers of components.

2.1.6 Fringe fields and edge focusing in dipole magnets

There is one further effect in dipoles that we have not so far discussed—the impact of fringe fields. In writing down the transfer matrices for dipoles and quadrupoles, we have assumed that the magnetic field rises over an infinitesimal distance from zero outside the magnet to its full strength within the magnet. However, in practice the change in the field cannot happen over an infinitesimal distance, but takes place gradually over a distance comparable with the transverse aperture of the magnet, which may be several centimetres. The region at the start (and end) of a magnet where the field is rising from (or falling to) zero is known as the *fringe field region*. The exact behaviour of the magnetic field in a fringe field region can be complicated, and depends on the type of magnet considered [10]. However, in a first analysis of particle motion in an accelerator, it is usually only necessary to consider fringe fields in dipoles, fringe fields in quadrupoles can usually be ignored.

At the end of a dipole magnet, the field lines tend to curve out, resulting in a longitudinal component in the field, i.e. a component parallel to the reference trajectory (see the diagram on the right in figure 2.5). Because of the shape of the field, the strength of the longitudinal component increases with vertical distance from the mid-plane of the magnet. If the dipole is constructed so that it has a geometry fitting within a sector of a circle, so that either end of the magnet is perpendicular to the reference trajectory (as shown in the diagram on the left in

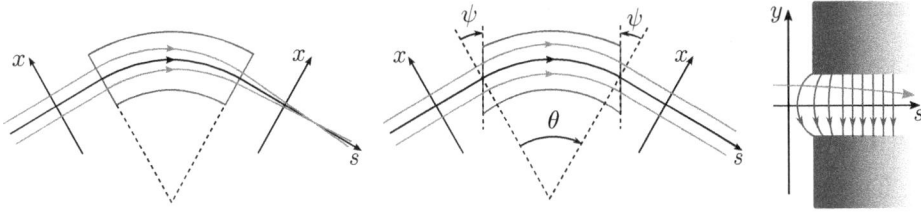

Figure 2.5. Edge focusing in a dipole magnet. If the pole faces in the dipole magnet are perpendicular to the reference trajectory (s-axis) as shown in the diagram on the left, then particles moving parallel to the reference trajectory at the entrance to the magnet experience weak focusing in the body of the magnet, so that the trajectories then converge to a focal point. If the pole faces are rotated by half of the bend angle, so $\psi = \theta/2$ in the central diagram, then each trajectory is simply a displacement in the horizontal plane of the reference trajectory—there is an effective 'edge focusing' (or defocusing, in this case) in the horizontal plane, that cancels the weak focusing from the body of the magnet. In the vertical plane, the rotation of the pole faces leads to particles crossing the fringe field (shown in a side view, in the diagram on the right) at an angle. Since the field lines in the fringe field curve out from the body of the magnet, they acquire a longitudinal component that increases with vertical displacement. A particle crossing the longitudinal component of the field at an angle experiences a vertical deflection. This effect leads to focusing in the vertical direction.

figure 2.5), then particles entering and leaving the magnet travelling parallel to the reference trajectory are also travelling parallel to the longitudinal component of the field, which therefore exerts no force on the particles. However, it is usual to 'rotate' the ends (pole faces) of a dipole magnet so that it has a rectangular footprint, as shown in the central diagram in figure 2.5. In that case, the fringe fields are also rotated, so that particles entering and leaving the magnet parallel to the reference trajectory cross the longitudinal component of the field at an angle. As a result, particles experience a vertical force, and since the longitudinal field component changes strength (and direction) with vertical position, the pole-face rotation in a dipole leads to vertical focusing.

Rotation of the pole faces of a dipole magnet also has an effect on the horizontal motion. If the pole faces are rotated so that the magnet has a rectangular footprint, then the distance over which particles experience the dipole field is reduced for particles on the outside of the arc defined by the reference trajectory, and increased for particles on the inside of this arc (see the central diagram in figure 2.5. This leads to an effective horizontal defocusing, resulting from the pole-face rotation. Rather curiously, the vertical focusing and horizontal defocusing are (in simple cases) matched, so that the overall effect from the fringe field with rotation of the pole faces is that of a quadrupole. In the 'thin lens' approximation (in which we take the length of the fringe field region to zero) the transfer matrix for a dipole fringe field can be written

$$
R_{\text{dipole,fringe}} = \begin{pmatrix} 1 & 0 & 0 & 0 & 0 & 0 \\ k_0 \tan(\psi) & 1 & 0 & 0 & 0 & 0 \\ 0 & 0 & 1 & 0 & 0 & 0 \\ 0 & 0 & -k_0 \tan(\psi) & 1 & 0 & 0 \\ 0 & 0 & 0 & 0 & 1 & 0 \\ 0 & 0 & 0 & 0 & 0 & 1 \end{pmatrix}, \tag{2.30}
$$

where $k_0 = qB/P_0$ with B the dipole field strength, and ψ is the rotation angle of the pole face of the dipole. If the pole face is angled so that it is perpendicular to the reference trajectory, then $\psi = 0$.

2.2 Betatron oscillations

We have seen in section 2.1.3 that in a quadrupole magnet, the vertical magnetic field varies linearly with horizontal distance from the axis of the magnet:

$$B_y = \frac{P_0}{q}k_1x, \tag{2.31}$$

where P_0/q is the ratio of the reference momentum to the particle charge, and k_1 is a constant, the (scaled) quadrupole field gradient. The horizontal force on a particle moving through this field is $F_x = -qvB_y$; therefore, using Newton's second law, the equation of motion for a particle moving through the quadrupole is

$$\frac{d^2x}{ds^2} = -k_1x, \tag{2.32}$$

where we assume that the particle has momentum close to the reference momentum. The equation of motion (2.32) is the equation for simple harmonic motion with wavelength $2\pi/\sqrt{k_1}$. We can generalise this result to other components in a storage ring. In a dipole, for example, the effect of weak focusing means that particles perform simple harmonic motion with wavelength $2\pi/\rho$, where ρ is the radius of curvature of the reference trajectory. In a drift space, there is no focusing, but we can still regard the particle as performing simple harmonic motion, though with infinite wavelength.

2.2.1 Hill's equation and the Courant–Snyder parameters

Since the transverse horizontal motion of a particle in an accelerator beamline (consisting of drift spaces, dipoles and quadrupoles) is essentially simple harmonic motion with varying wavelength, we can write the equation of motion in general as follows:

$$\frac{d^2x}{ds^2} = -k_1(s)x, \tag{2.33}$$

where the focusing strength $k_1(s)$ is a function of the position along the reference trajectory. The value of the focusing strength at a given position depends on the magnetic field (or the gradient of the magnetic field) present at that location.

In a storage ring, the focusing strength varies periodically—a particle sees the same sequence of components on each turn that it makes around the storage ring. When $k_1(s)$ is a periodic function, the equation of motion (2.33) is known as *Hill's equation* [11, 12]. The solution to Hill's equation can be written

$$x(s) = \sqrt{2J_x\beta_x(s)}\cos(\phi_x), \tag{2.34}$$

where J_x is a constant of integration, $\beta_x(s)$ is a periodic function with the same periodicity as $k_1(s)$, and the phase ϕ_x varies with position along the reference trajectory as

$$\frac{d\phi_x}{ds} = \frac{1}{\beta_x(s)}. \tag{2.35}$$

The function $\beta_x(s)$ is known (in accelerator physics) simply as the *beta function*, and turns out to play a central role in accelerator beam dynamics. From the solution to Hill's equation, we can see that it describes two features of the motion of a particle in a storage ring. First, from (2.34), we see that the motion resembles that of a simple harmonic oscillator, but with variation in the amplitude described by $\beta_x(s)$. Second, from (2.35), we see that the wavelength of the motion, corresponding to the distance over which the phase ϕ_x increases by 2π, is given by $2\pi\beta_x(s)$. These are key results: the motion of a particle around a storage ring looks like simple harmonic motion, though with amplitude and wavelength that vary with position in the ring, depending on the focusing strength at each point around the ring. These results can be extended to the vertical motion, although it should be remembered that a quadrupole that provides a focusing force (positive k_1) in the horizontal direction provides a defocusing force in the vertical direction; and that apart from the effects of fringe fields, there is no vertical focusing in a dipole magnet. The transverse horizontal and vertical oscillations of particles in a storage ring are known as *betatron oscillations*. The quantity J_x is the (horizontal) betatron amplitude (also known as the *betatron action*), and ϕ_x is the (horizontal) betatron phase.

Since the beta function plays a central role in describing particle motion in a storage ring, it is important to be able to calculate its value at any point in a storage ring with a given sequence of components. Before we discuss how to work out the value of the beta function, let us make an apparent digression to consider the idea of a *matched distribution*; it turns out that this discussion will be useful in finding a way to calculate the beta function.

The size of a beam of particles in a storage ring can be characterised in terms of the second-order moments of the particle distribution. Consider a beam of particles passing a given point along the reference trajectory. Suppose we record the horizontal co-ordinate of each particle in the beam as it passes this point; the second-order moment of the particle co-ordinates is the square of the horizontal co-ordinate, averaged over all the particles in the beam. We write this as $\langle x^2 \rangle$. Assuming that we can also measure the horizontal momentum of each particle, we can also calculate the *divergence* of the beam at the given point, $\langle p_x^2 \rangle$ and the correlation between the co-ordinates and momenta $\langle x p_x \rangle$.

The motion of each particle in a storage ring must be described by the same beta function. However, at any given location around the ring, each particle can have its own action J_x and phase ϕ_x. From the solution to Hill's equation (2.34), we then find:

$$\langle x^2 \rangle = 2\beta_x \langle J_x \cos^2(\phi_x) \rangle, \tag{2.36}$$

where to simplify the notation, we do not show explicitly that β_x is a function of s. Let us assume that the betatron amplitudes of the different particles in the beam are not correlated with the betatron phases, and that the betatron phases are randomly and evenly distributed between 0 and 2π. With these assumptions, we can replace the function $\cos^2(\phi_x)$ in the expression (2.36) for $\langle x^2 \rangle$ by its average value of $\frac{1}{2}$, and we obtain:

$$\langle x^2 \rangle = \beta_x(s)\langle J_x \rangle. \tag{2.37}$$

We can carry out a similar process to find expressions for the beam divergence $\langle p_x^2 \rangle$ and the correlation $\langle xp_x \rangle$. First of all, since $p_x \approx dx/ds$, by taking the derivative of the solution to Hill's equation (2.34) we find that:

$$p_x = -\sqrt{\frac{2J_x}{\beta_x}}\,(\sin(\phi_x) + \alpha_x \cos(\phi_x)), \tag{2.38}$$

where we define the *alpha function*

$$\alpha_x(s) = -\frac{1}{2}\frac{d\beta_x}{ds}. \tag{2.39}$$

Then, with the same assumptions that we used when calculating $\langle x^2 \rangle$, we find that

$$\langle xp_x \rangle = -\alpha_x \langle J_x \rangle, \tag{2.40}$$

$$\left\langle p_x^2 \right\rangle = \gamma_x \langle J_x \rangle. \tag{2.41}$$

The *gamma function* $\gamma_x(s)$ (not to be confused with the relativistic factor, γ) is defined so that

$$\beta_x(s)\gamma_x(s) - \alpha_x(s)^2 = 1. \tag{2.42}$$

The beta, alpha, and gamma functions $\beta_x(s)$, $\alpha_x(s)$ and $\gamma_x(s)$ are known as the *Courant–Snyder parameters*[4] [14].

2.2.2 The matched distribution in a periodic lattice

Now suppose that we take a bunch of particles at a given point along the reference trajectory in a storage ring, and transport the bunch around one complete turn of the ring. Since the horizontal co-ordinate and momentum will change after making the revolution of the ring, we might expect that the second-order moments of the particle distribution will also change. In an electron storage ring, however, we observe that the beam quickly finds an equilibrium distribution that remains constant in time at a given point in the ring. This implies the existence of a *matched distribution* that stays the same after all the particles are transported once around the

---

[4] The beta, alpha and gamma functions are also often referred to as the *Twiss parameters* [13].

ring, even though the horizontal and vertical co-ordinates and momenta of each particle change.

It is helpful to introduce at this point the *covariance matrix*, which is a matrix constructed from the second-order moments of the particle distribution:

$$\Sigma(s) = \begin{pmatrix} \langle x^2 \rangle & \langle xp_x \rangle \\ \langle xp_x \rangle & \langle p_x^2 \rangle \end{pmatrix} = B(s)\langle J_x \rangle. \tag{2.43}$$

The elements of the matrix $B(s)$ are the Courant–Snyder parameters at the given position s along the reference trajectory:

$$B(s) = \begin{pmatrix} \beta_x(s) & -\alpha_x(s) \\ -\alpha_x(s) & \gamma_x(s) \end{pmatrix}. \tag{2.44}$$

Since the transfer matrices for the individual components in a storage ring are independent of the particle co-ordinates and momenta, we can define the *single-turn transfer matrix M* as the product of the transfer matrices of each component in the ring in sequence:

$$M = R_n R_{n-1} \cdots R_2 R_1. \tag{2.45}$$

Then, given the initial horizontal co-ordinate and momentum of any particle, we can calculate the co-ordinate and momentum of the particle after one turn around the storage ring using the single-turn transfer matrix:

$$\begin{pmatrix} x \\ p_x \end{pmatrix}_{s_0+C_0} = M \begin{pmatrix} x \\ p_x \end{pmatrix}_{s_0}. \tag{2.46}$$

From the change in the co-ordinate and momentum of each particle, we can calculate the change in the covariance matrix over one turn. Since the covariance matrix involves second-order moments of the co-ordinates and momenta, we need to apply two copies of the single-turn transfer matrix:

$$\Sigma(s_0 + C_0) = M\Sigma(s_0)M^{\mathrm{T}}, \tag{2.47}$$

where M^T is the transpose of the single-turn transfer matrix. Suppose that we start with a matched distribution, described (in terms of the second-order moments of the particle distribution) by particular values of the Courant–Snyder parameters:

$$\Sigma_{\text{matched}}(s_0) = B(s_0)\langle J_x \rangle. \tag{2.48}$$

By definition, the matched distribution has the property that it remains the same after one complete turn around the ring, and so

$$M\Sigma_{\text{matched}}(s_0)M^{\mathrm{T}} = \Sigma_{\text{matched}}(s_0). \tag{2.49}$$

It can be verified (by multiplying the matrices) that this equation is satisfied if the single-turn transfer matrix has the form

$$M = I \cos(\mu_x) + B(s_0)S \sin(\mu_x), \tag{2.50}$$

where I is the identity matrix and S is the matrix

$$S = \begin{pmatrix} 0 & 1 \\ -1 & 0 \end{pmatrix}. \tag{2.51}$$

2.2.3 Betatron phase advance and the betatron tunes

The parameter μ_x in (2.50) is associated with the change in the betatron phase ϕ_x when the transfer matrix M is applied to the co-ordinate x and momentum p_x of a given particle. Suppose that we take the transfer matrix in terms of μ_x and the Courant–Snyder parameters (2.50) and apply it to a given co-ordinate (2.34) and momentum (2.38) expressed in terms of the Courant–Snyder parameters, betatron amplitude J_x and betatron phase ϕ_x. We then find that the new co-ordinate and momentum are described by the same Courant–Snyder parameters and betatron amplitude as the original co-ordinate and momentum, but that the betatron phase has increased by μ_x; that is, $\phi_x(s_0 + C_0) = \phi_x(s_0) + \mu_x$. In other words, μ_x is the *phase advance* associated with the transfer matrix M. In a storage ring, we define the *betatron tune* ν_x as the number of betatron oscillations made by a particle in one turn of a storage ring. The phase advance is then given by $\mu_x = 2\pi \nu_x$.

For a given sequence of components in the storage ring, we can construct the single-turn transfer matrix M, by multiplying the transfer matrices for the individual components. Given the elements of the transfer matrix found in this way from the design of the storage ring, we can solve equation (2.50) to find the Courant–Snyder parameters (contained in the matrix $B(s_0)$) that describe the matched distribution, and the phase advance μ_x. This is reasonably straightforward. First, the phase advance μ_x can be found from

$$2\cos(\mu_x) = \mathrm{Tr}(M), \tag{2.52}$$

where $\mathrm{Tr}(M)$ is the trace of the matrix M. Strictly speaking, this gives the phase advance modulo 2π, but that is sufficient for our present purposes. Then the matrix $B(s_0)$ is given by

$$B(s_0) = \frac{S}{\sin(\mu_x)}(I \cos(\mu_x) - M). \tag{2.53}$$

Notice that the Courant–Snyder parameters are undefined if the betatron tune is an integer (so that $\mu_x = 0$) or a half-integer (so that $\mu_x = \pi$). For an integer tune, the co-ordinate and momentum of any particle remain unchanged after one complete turn through the storage ring; in that case, resonance can occur between the particle motion and the deflection that the particle receives in dipole magnets, resulting in the particle motion becoming unstable. For a half-integer tune, resonance can again occur, but driven this time by the quadrupole magnets. We shall return to the subject of resonances in section 4.2.

In addition to the possibility of resonances, we see from (2.52) that there is a condition on the trace of the transfer matrix for the motion of particles in a storage ring to be stable. Specifically, if the trace of the transfer matrix has a magnitude larger than 2, then the phase advance is an imaginary number. In that case, the motion of a particle must be expressed in terms of hyperbolic functions, rather than regular trigonometric functions. This means that the particle does not oscillate around the reference trajectory as it moves round the storage ring; instead, the horizontal co-ordinate (and momentum) will increase exponentially, until the particle hits the wall of the vacuum chamber and is lost from the beam.

It is worth noting that the procedure described above for finding the Courant–Snyder parameters for a matched distribution can be applied to any periodic beamline, not just a storage ring. Accelerator beamlines are often constructed from repeated units, called 'cells', of a given sequence of components; using the transfer matrix M for a single cell in equation (2.50), it is possible to find the Courant–Snyder parameters that describe a particle distribution that remains the same from one cell to the next.

We also note that given the Courant–Snyder parameters at any point in a beamline, we can find the Courant–Snyder parameters at any other point from the transfer matrix between those two points. The matrix B is related directly to the covariance matrix (2.43), and so must transform in the same way as the covariance matrix when particles are transported from one point to another. Thus

$$B(s_1) = R(s_1, s_0)B(s_0)R(s_1, s_0)^\mathrm{T}, \qquad (2.54)$$

where $R(s_1, s_0)$ is the transfer matrix from $s = s_0$ to $s = s_1$. Using this relationship, once we have found the Courant–Snyder parameters at any point in a storage ring (or any periodic beamline), the parameters at any other point can be found relatively easily.

2.2.4 Action–angle variables

We saw above that the co-ordinate x and momentum p_x of a particle can be expressed in terms of the Courant–Snyder parameters, the betatron amplitude J_x and the betatron phase ϕ_x. The Courant–Snyder parameters are a property of the beamline, and so are the same (at a given point in a beamline) for all the particles in a beam. However, each particle will, in general, have its own amplitude and phase. Given a set of Courant–Snyder parameters, the quantities J_x and ϕ_x provide an alternative to the variables x and p_x for describing the motion of a particle along a beamline. The amplitude J_x and phase ϕ_x are known in classical mechanics as *action–angle variables* [7]. The advantage of using action–angle variables is that the amplitude (or action) J_x is a constant for a given particle, as that particle moves along a beam line. This is apparent from the fact that the action was introduced as a constant of integration in the solution of Hill's equation (2.34).

An illustration of the physical significance of the action–angle variables can be found by plotting the co-ordinate x and momentum p_x for a given value of the action J_x and for the angle (betatron phase) varying over the range 0 to 2π. The result is an ellipse in phase space, with an area $2\pi J_x$, and with a shape described by the Courant–Snyder parameters. Such an ellipse could be constructed, for example, by plotting the co-ordinate and momentum of a particle tracked through successive cells in a periodic lattice (see figure 2.6).

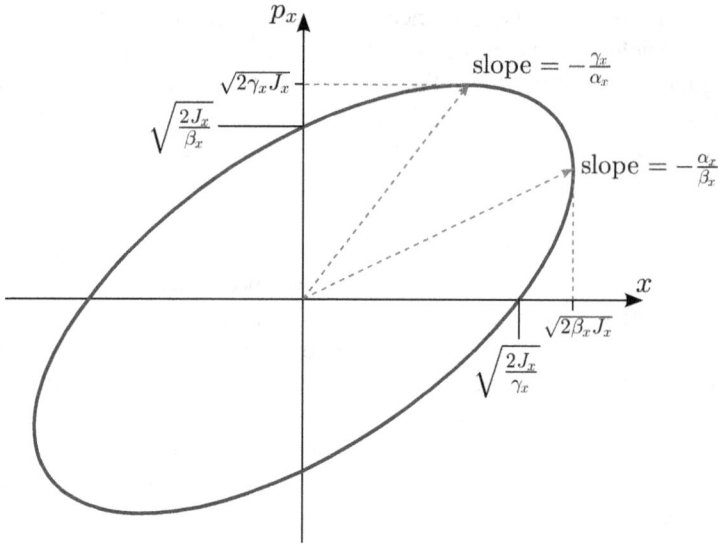

Figure 2.6. Phase space trajectory for a particle performing betatron oscillations in a periodic lattice. If the co-ordinate x and momentum p_x of the particle are plotted in phase space after each pass through one periodic section of the lattice, the points lie on an ellipse with area $2\pi J_x$, where J_x is the betatron action. The betatron phase ϕ_x gives the position of the particle around the ellipse. The shape of the ellipse is determined by the Courant–Snyder parameters β_x, α_x, and γ_x.

It is sometimes useful to express the betatron action and betatron phase in terms of the co-ordinate and momentum of a particle. By inverting the relationships already given, (2.34) and (2.38), we find

$$2J_x = \gamma_x x^2 + 2\alpha_x x p_x + \beta_x p_x^2, \tag{2.55}$$

$$\tan(\phi_x) = -\beta_x \frac{p_x}{x} - \alpha_x. \tag{2.56}$$

With the relationship between the Courant–Snyder parameters (2.42), equation (2.55) is the equation for an ellipse with area $2\pi J_x$.

2.3 Emittance

In section 2.2 we saw that the second-order moments of the particle distribution in an accelerator beamline could be written in terms of the Courant–Snyder parameters and the mean betatron amplitude. For convenience, we define the *transverse emittance* of a beam as the average betatron amplitude of all particles in the beam; thus, the horizontal emittance is given by $\epsilon_x = \langle J_x \rangle$, and similarly for the vertical emittance. The second-order moments of the beam distribution can then be written

$$\langle x^2 \rangle = \beta_x \epsilon_x, \tag{2.57}$$

$$\langle xp_x \rangle = -\alpha_x \epsilon_x, \tag{2.58}$$

$$\left\langle p_x^2 \right\rangle = \gamma_x \epsilon_x. \tag{2.59}$$

Since the action of each particle is a constant of the motion for each individual particle (i.e. the action remains the same as the particle moves around the storage ring), it follows that the emittance is also a constant of the motion. In fact, it can be shown that as a bunch moves around a storage ring, the particle density in *phase space* in the vicinity of any given particle in the bunch remains constant—this is known as *Liouville's theorem*. If we consider only horizontal motion, a point in phase space is specified by the values of the co-ordinate x and the momentum p_x; in general, a point in phase space is specified by the co-ordinates x, y, z, the momenta p_x, p_y, and the energy deviation[5] δ.

Liouville's theorem is valid if we neglect synchrotron radiation, and a number of other effects that can change the betatron amplitudes of particles in an accelerator. We shall see in chapter 3 that in electron storage rings, synchrotron radiation leads to a beam reaching an equilibrium horizontal emittance that depends on the design of the storage ring lattice. There is also an equilibrium vertical emittance, although this is generally determined in practice by the alignment and tuning of the storage ring components, rather than by the design of the lattice.

It is often convenient to be able to calculate the emittance of a beam from the second-order moments of the particle distribution. Since the Courant–Snyder parameters satisfy the relationship $\beta_x \gamma_x - \alpha_x^2 = 1$, we find from the expressions (2.57), (2.58), and (2.59) given above, that:

$$\epsilon_x = \sqrt{\langle x^2 \rangle \left\langle p_x^2 \right\rangle - \langle xp_x \rangle^2}. \tag{2.60}$$

The emittance ϵ_x defined above is sometimes called the *geometric emittance* to distinguish it from the *normalised emittance* that is also used in accelerator physics. The normalised emittance is defined as $\epsilon_{x,n} = \gamma \epsilon_x$, where γ is the relativistic factor $1/\sqrt{1 - v^2/c^2}$ for a particle with velocity v (and not to be confused with the Courant–Snyder parameter γ_x). The normalised emittance is commonly used when discussing beam dynamics in an accelerator in which the particles experience a change in energy (for example, in a linac), because in that case it is the normalised emittance that is a constant of the motion, rather than the geometric emittance. Since the normalised emittance remains constant if the energies of the particles (and hence the relativistic factor γ) increase, this implies that the geometric emittance is reduced when a beam is accelerated: this effect is known as *adiabatic damping*. In a storage ring, if the beam is held at a fixed energy (neglecting the small changes arising from synchrotron radiation and the RF cavities), there is no adiabatic damping, and it is

---

[5] It should be remembered that the z co-ordinate specifies the *time* at which a particle arrives at a given point in the beamline, relative to the reference particle (2.17). The energy deviation is defined by (2.18).

then conventional to use the geometric emittance in discussions of the beam dynamics, rather than the normalised emittance.

2.4 The closed orbit

When we introduced the reference trajectory in section 2.1, we emphasised the fact that we could define the reference trajectory for our own convenience in specifying the origin of the transverse co-ordinate system (with axes x and y) at any position along an accelerator beamline. The reference trajectory does not need to be a possible physical trajectory for a particle moving along the beamline. In a storage ring, however, it is usual to define the reference trajectory so that a particle starting on the reference trajectory will continue to follow the reference trajectory over one complete turn of the ring, and will return to its original position at the end of the turn. The reference trajectory in this case forms a *closed orbit*, which is simply a trajectory in a storage ring that can be followed by a real physical particle (obeying the appropriate equations of motion determined by the electromagnetic fields around the ring) and that returns to the original starting point after one complete turn around the ring.

Suppose that in an otherwise 'perfect' lattice, we introduce a small error on one of the magnets. For example, the current in a dipole magnet is reduced slightly, leading to a reduction in the field strength compared to the specified design strength. A particle moving through the dipole will no longer follow the (design) reference trajectory, since the radius of curvature of the particle trajectory in the reduced dipole field will be larger than that of the reference trajectory. However, although errors in field strength and alignment will change the trajectories of particles in a storage ring, it may still be possible to find a closed orbit in the presence of such errors. Assuming that a closed orbit exists, it can be calculated from the single-turn transfer map, as we now describe.

For a lattice consisting of drift spaces, dipole magnets, quadrupole magnets and RF cavities, we can write the effect of the single-turn transfer map for particles in a storage ring as follows:

$$\begin{pmatrix} x \\ p_x \end{pmatrix}_{s_0 + C_0} = M \begin{pmatrix} x \\ p_x \end{pmatrix}_{s_0} + \vec{m}, \tag{2.61}$$

where M is the single-turn transfer matrix, and $\vec{m}$ is a two-component vector with constant elements. If we know the parameters of all the ring components, including all the alignment and field strength errors, then we can calculate the transfer matrix M and the vector $\vec{m}$. By definition, the co-ordinate and momentum of a particle on the closed orbit satisfy the condition

$$\begin{pmatrix} x_{co} \\ p_{x,co} \end{pmatrix}_{s_0} = M \begin{pmatrix} x_{co} \\ p_{x,co} \end{pmatrix}_{s_0} + \vec{m}, \tag{2.62}$$

which simply states that after a complete turn of the ring, the co-ordinate and momentum of the particle return to their original values. Rearranging this equation gives

$$\begin{pmatrix} x_{\text{co}} \\ p_{x,\text{co}} \end{pmatrix}_{s_0} = (I - M)^{-1}\vec{m}, \qquad (2.63)$$

where I is the identity matrix, and $(I - M)^{-1}$ is the inverse of $I - M$.

There is no guarantee that the matrix $I - M$ can be inverted; but if it can, then equation (2.63) can be used to find the closed orbit. If the matrix $I - M$ cannot be inverted (if it has determinant equal to zero, or very close to zero) then the closed orbit does not exist, or may be so far from the reference trajectory that in places it passes out of the vacuum chamber—it will not be possible under such conditions to store beam in the storage ring.

2.5 Dispersion

We saw in section 2.1.2 that the trajectory of a particle through a dipole magnet depends on the energy of the particle. The variation in trajectory resulting from a change in energy is known as *dispersion*. More precisely, the dispersion η_x can be defined as the rate of change of the trajectory with respect to the energy deviation. In a storage ring, the appropriate trajectory to consider is the closed orbit (section 2.4):

$$\eta_x = \frac{dx_{\text{co}}}{d\delta}, \qquad (2.64)$$

where the derivative should be evaluated at $\delta = 0$ to avoid any impact from nonlinearities in the lattice (arising, for example, from sextupole magnets).

Measurement of the dispersion in a storage ring is a routine procedure used to characterise the optics, and to assess the impact of errors in the alignment and strength of the beamline components. The procedure involves changing the beam energy by a small amount, and observing the resulting change in the closed orbit from beam position monitors (BPMs) distributed around the ring. Although it might be expected that a change in the beam energy could be achieved by changing the voltage of the RF cavities in the ring (since the change in energy of a particle passing through the cavity depends on the voltage across the cavity), in fact changing the voltage has no impact on the beam energy. This is because particles in the beam find a new equilibrium in which they change their arrival time at the cavity, with respect to the phase of the RF field. As a result, although the amplitude V_0 of the voltage may have changed, the voltage observed by the beam $V_0 \sin(\phi_s)$ remains the same. However, if the voltage is kept constant but the RF frequency is changed, then in order to maintain the synchronisation between the revolution frequency and the RF frequency, the closed orbit has to change (either shrink or expand in circumference). The change in the closed orbit is achieved by a change in the beam energy, combined with the effect of dispersion.

A calculation of the dispersion in a computer model of a storage ring can be achieved in a similar way to the measurement of the dispersion in a real storage ring. First, the closed orbit in the ring under the nominal conditions is calculated, using the procedure described in section 2.4, but extended to include the longitudinal variables (z, δ) as well as the horizontal variables (x, p_x) (so that M is now a

4×4 matrix). Then, a small 'circumference error' is introduced by modifying the vector $\vec{m}$ as follows:

$$\vec{m}_\Delta = \vec{m} + \begin{pmatrix} 0 \\ 0 \\ \Delta z \\ 0 \end{pmatrix}. \qquad (2.65)$$

The new closed orbit is calculated using

$$\begin{pmatrix} x_\Delta \\ p_{x,\Delta} \\ z_\Delta \\ \delta_\Delta \end{pmatrix}_{s_0} = (I - M)^{-1}\vec{m}_\Delta. \qquad (2.66)$$

The dispersion is found from

$$\eta_x(s_0) = \lim_{\Delta z \to 0} \frac{x_\Delta}{\delta_\Delta}. \qquad (2.67)$$

In a planar storage ring, in which the closed orbit lies entirely in a horizontal plane, there is no vertical bending, and therefore the vertical dispersion is zero. However, errors in the alignment of the magnets (for example, a vertical displacement of a quadrupole magnet, or a tilt of a dipole magnet so that the field is no longer perfectly vertical) can distort the vertical closed orbit, leading to vertical dispersion. The vertical dispersion in that case can be calculated (and measured) in the same way as the horizontal dispersion.

Vertical dispersion can also be generated by coupling errors, which result in the vertical motion of a particle depending directly on the horizontal motion, and vice versa. Coupling is discussed in more detail in section 2.6; but for now, we note that to calculate the (horizontal and vertical) dispersion in a model of a storage ring, the method described above can be generalised to use the full 6×6 transfer matrices, rather than 4×4 transfer matrices that deal with the horizontal and vertical motion separately.

Dispersion (horizontal or vertical) will contribute to the size of a beam observed in an accelerator beam line, because where dispersion is present the transverse position of a particle will depend on its energy as well as on the betatron action. This means that the horizontal co-ordinate of a particle (2.34) should be written (for a particle with non-zero energy deviation):

$$x(s) = \sqrt{2J_x\beta_x} \cos(\phi_x) + \eta_x\delta. \qquad (2.68)$$

Assuming that the energy deviation is not correlated with the betatron phase, the mean-square beam size (2.57) then becomes

$$\langle x^2 \rangle = \beta_x\epsilon_x + \eta_x^2\sigma_\delta^2, \qquad (2.69)$$

where $\sigma_\delta = \sqrt{\langle \delta^2 \rangle}$ is the rms energy spread. In an electron storage ring, a typical value for the equilibrium rms energy spread (determined by synchrotron radiation effects, as discussed in section 3.2) is 0.1%, and is roughly constant around the full circumference. If the horizontal emittance is 5 nm, then at a location in the ring where the beta function is 10 m and the dispersion is 0.2 m (using typical values) the horizontal rms beam size will be $\sigma_x = \sqrt{\langle x^2 \rangle} = 300$ μm, with the emittance and the energy spread making roughly equal contributions.

Taking account of dispersion, the correlation $\langle xp_x \rangle$ (2.58) and the beam divergence $\langle p_x^2 \rangle$ (2.59) become

$$\langle xp_x \rangle = -\alpha_x \epsilon_x + \eta_x \eta_x' \sigma_\delta^2, \tag{2.70}$$

$$\left\langle p_x^2 \right\rangle = \gamma_x \epsilon_x + \eta_x'^2 \sigma_\delta^2, \tag{2.71}$$

where η_x' is the gradient of the dispersion along the reference trajectory, i.e. $\eta_x' = d\eta_x / ds$.

2.6 Coupling

Coupling describes how the motion of a particle in one direction (horizontal, vertical or longitudinal) depends on the motion in another direction. In sections 2.1 and 2.2 we assumed that the vertical motion could be treated completely separately from the horizontal motion—in other words, we assumed that there was no *betatron coupling*. In a storage ring consisting entirely of drift spaces, dipole magnets, quadrupole magnets, and RF cavities, this is a valid assumption, as long as the magnets are properly aligned (for instance, so that the horizontal field component in each quadrupole magnet is zero in the plane of the reference trajectory). However, in a storage ring there is always some dispersion generated by the dipole magnets. This is a form of coupling, because the horizontal motion then depends on the energy deviation δ (a longitudinal variable), and the revolution period (described by changes in the longitudinal co-ordinate z) depends on the horizontal motion.

In practice, because magnets in a storage ring can never be perfectly aligned, betatron coupling (coupling between the horizontal and vertical motion) is always present to some extent. Consider, for example, a quadrupole magnet that is rotated (tilted) slightly around the magnetic axis. This introduces a *skew quadrupole* component in the magnetic field. A particle travelling with some horizontal offset x from the reference trajectory then sees a horizontal component of the magnetic field that increases with increasing x. There is, as a result, a vertical force on the particle that depends on its horizontal position. Similarly, there will be a horizontal force on the particle that depends on its vertical position. In other words, there will be some betatron coupling in the storage ring.

The motion of particles in the presence of betatron coupling becomes considerably more complicated than in the case without betatron coupling. Coupling can be readily accommodated in the 6×6 transfer matrices that we developed in section 2.1, but the Courant–Snyder parameters can be generalised in a number of different ways to take account of betatron coupling (see, for example, [15–17]), and no single method has been universally adopted. Largely, this is because betatron coupling has a number of different effects on the beam behaviour in a storage ring, and the most appropriate method for analysing the dynamics in the presence of coupling depends on the effect being considered.

One way in which betatron coupling can impact the performance of an electron storage ring is by increasing the vertical emittance. As we shall see in chapter 3, synchrotron radiation in an electron storage ring leads to a natural emittance ϵ_0, which is shared between the horizontal and vertical motion of the particles. In the absence of betatron coupling, the natural emittance contributes entirely to the horizontal motion, so that the horizontal emittance is equal to the natural emittance, and the vertical emittance is close to zero[6]. If betatron coupling is very strong, then the natural emittance is shared equally between the horizontal and vertical motion, so that the horizontal and vertical emittances become (roughly) the same.

Measuring and correcting the coupling is often an important task in the operation of an electron storage ring, because the vertical emittance is a key figure of merit for many applications. For example, in a synchrotron light source the photon beam brightness depends directly on the vertical (as well as on the horizontal) emittance of the electron beam. Similarly, in a collider the luminosity depends on the vertical emittances of the colliding beams. A common method of characterising the coupling in a storage ring is by measuring the smallest 'tune split' that can be achieved by adjusting the quadrupole strengths. The tune split is simply the difference between the horizontal and vertical tunes. Since the tunes depend on the quadrupole strengths in a storage ring, by adjusting the strengths appropriately it should be possible (in principle) to achieve conditions where the horizontal and vertical tunes are exactly equal. However, the presence of betatron coupling affects not just the emittances, but also the tunes. It turns out that betatron coupling prevents the tunes becoming exactly equal. In fact, the smallest difference that can be achieved between the tunes provides a useful measure of the amount of coupling (i.e. the strength of the magnetic fields around the storage ring that cause the coupling)—see, for example [18]. Control of the betatron coupling can be achieved using skew quadrupoles placed at selected locations around the storage ring. By setting the strengths of the skew quadrupoles appropriately, it is often possible to cancel the effects of tilts of the normal quadrupoles. The optimum strengths of the skew quadrupoles used for correcting the coupling can be found empirically, by observing the effect of changing their strengths on the minimum tune split; but various techniques have been developed to achieve good control of the coupling in electron storage rings [19].

In practice, storage rings in synchrotron light sources often aim to correct the coupling to a level that provides a vertical emittance that is of order 1% of the

---

[6] There are some effects (in particular, the emission of synchrotron radiation at an angle to the horizontal plane) that lead to a small but non-zero value of the vertical emittance even in the absence of betatron coupling and vertical dispersion; this is discussed further in section 3.2.3.

horizontal emittance. Without correction of the coupling, the vertical emittance could easily be around 50% (or more) of the horizontal emittance. Reducing the vertical emittance below 1% of the horizontal may provide some increase in brightness of the photon beams produced by synchrotron radiation, but may not benefit the operation overall, since there are also adverse effects (for example, Touschek scattering, which limits the beam lifetime; see section 5.1) that become more severe as the vertical emittance is reduced. As is often the case, the optimum operational conditions are a compromise between a number of competing effects, and depend very much on the requirements of the user experiments, which can change over time. The real goal is to achieve control over all the various aspects of the storage ring operation (including the betatron coupling), so as to be able to provide the best operational conditions for the user experiments at any given time.

2.7 Momentum compaction factor

The *momentum compaction factor* α_p describes the change in the length of the closed orbit with respect to a change in particle energy:

$$\alpha_p = \frac{1}{C_0} \frac{dC}{d\delta}, \tag{2.72}$$

where C is the length of the closed orbit over one turn of the storage ring, δ is the energy deviation, and C_0 is the length of the closed orbit for a particle with zero energy deviation. As we discussed in section 2.5, the closed orbit depends on the energy of a particle—the change in the closed orbit with respect to the energy deviation is described by the dispersion. The change in the circumference of the closed orbit with respect to the particle energy is an important quantity in the longitudinal motion of particles in a storage ring. As we shall see in chapter 3, the momentum compaction factor appears (for example) in the expression for the equilibrium bunch length determined by synchrotron radiation effects, and this has implications for the amount of power that must be supplied to the RF cavities in a storage ring, and for the beam lifetime.

In a section of a storage ring with a straight reference trajectory, any change in the particle trajectory generally has a small impact on the length of the reference trajectory. In particular, if a particle is travelling parallel to the reference trajectory, the length of the trajectory will be the same no matter what the transverse offset happens to be. However, where the reference trajectory is curved (which is usually the case in a dipole magnet) the length of the particle trajectory depends on the distance (in the plane of the curvature) of the trajectory from the reference trajectory. If the reference trajectory has radius of curvature ρ, then a distance ds along the reference trajectory subtends an angle $\theta = ds/\rho$ at the centre of curvature. A particle with constant offset x from the reference trajectory follows a trajectory with radius of curvature $\rho + x$; as we see from figure 2.7, over the same section of the accelerator (i.e. subtending the same angle θ at the centre of curvature) the length of

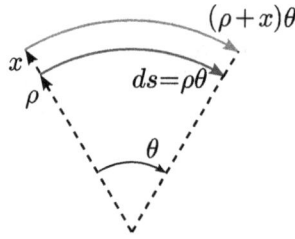

Figure 2.7. Change in path length for a particle in a dipole magnet (with bending radius ρ) resulting from the horizontal displacement of the particle from the reference trajectory. A particle on the reference trajectory (blue arc) follows a path of length $ds = \rho\theta$ where θ is the angle subtended by the path at the centre of curvature. A particle with horizontal offset x from the reference trajectory follows a path (red arc) of length $(\rho + x)\theta$.

the trajectory followed by the particle is $(\rho + x)\theta = (\rho + x)ds/\rho$. The change in the length of the trajectory (with respect to the reference trajectory) is then

$$\Delta C = \frac{x}{\rho}ds. \tag{2.73}$$

To calculate the momentum compaction factor, we have to add up the change in the length of the particle trajectory over the entire circumference of the storage ring, assuming that the change in trajectory comes from the change in energy of a particle. By definition of the dispersion, we can write

$$x = \eta_x \delta. \tag{2.74}$$

Therefore, using the above result (2.73) and integrating the change in length of the trajectory around the entire ring we find

$$C = C_0 + \int_0^{C_0} \frac{\eta_x \delta}{\rho}ds, \tag{2.75}$$

where C_0 is the length of the closed orbit for a particle with $\delta = 0$. Finally, from definition of the momentum compaction factor (2.72), we have

$$\alpha_p = \frac{1}{C_0} \int_0^{C_0} \frac{\eta_x}{\rho}ds. \tag{2.76}$$

A particle in a uniform magnetic field follows a circular (or helical) trajectory— the radius of curvature increases if the energy of the particle is increased. This means we generally expect the circumference of the closed orbit to increase with the energy deviation, so the momentum compaction factor is usually a positive quantity. For a particle moving in a uniform magnetic field, $\alpha_p = 1$. However, the dispersion is normally controlled using quadrupole magnets, to achieve particular machine parameters (for example, a specified natural emittance), and in a storage ring designed for a low-emittance light source, the value of the momentum compaction factor may be of order 10^{-4}. It is even possible in some cases to tune the quadrupole magnets to achieve values of the momentum compaction factor that are negative, although this is not usually done on a routine basis.

A quantity related to the momentum compaction factor is the *phase slip factor*, η_p. The phase slip factor describes the change in the revolution period with respect to particle energy:

$$\eta_p = \frac{1}{T_0}\frac{dT}{d\delta}, \tag{2.77}$$

where T is the revolution period, i.e. the time taken for a particle to travel once around the ring on the closed orbit, and T_0 is the revolution period for a particle with zero energy deviation. Since the revolution period is simply the circumference divided by the velocity of the particle, the momentum compaction factor and the phase slip factor are related by the energy of the particle. In fact, it can be shown that

$$\eta_p = \alpha_p - \frac{1}{\gamma_0^2}, \tag{2.78}$$

where γ_0 is the relativistic gamma factor for a particle with zero energy deviation.

The relationship (2.78) between the phase slip factor and the momentum compaction factor provides a nice illustration of relativistic effects in storage rings. Suppose that a beam is stored in a ring at a low energy, so that $1/\gamma_0^2$ is larger than the momentum compaction factor. In that case, the phase slip factor will be negative, meaning that an increase in particle energy results in a reduction of the revolution period. This is the situation we would expect for sub-relativistic particles. Although the circumference of the closed orbit increases with increasing energy, the particle velocity also increases, and more than compensates (in terms of the revolution period) the increase in circumference. However, if the beam energy is high enough so that $1/\gamma_0^2$ is less than the momentum compaction factor, then the phase slip factor is positive. An increase in particle energy leads to an increase in circumference as before; but because the particles are already travelling at close to the speed of light there is insufficient change in the particle velocity to compensate the increase in circumference. As a result, the revolution period increases with increasing energy.

The beam energy at which an increase in circumference of the closed orbit is exactly balanced by an increase in particle velocity is known as the *transition energy*. At the transition energy, the phase slip factor is zero: this tends to have adverse consequences for the longitudinal dynamics. If a beam is injected into a ring at low energy (below transition), and the energy is then increased to a point above transition (by increasing the strengths of the fields in the magnets), special consideration needs to be given to passing through the transition energy in order not to lose the beam from the ring. Fortunately, electron storage rings tend to operate purely above transition, even at relatively low energy. This is in contrast to proton storage rings: because the proton mass is much larger than that of the electron, the relativistic factor for electrons at a given energy will be much larger than that for protons at the same energy. It is not uncommon for proton storage rings to operate below transition.

2.8 Synchrotron oscillations and phase stability

In section 2.2 we discussed the betatron oscillations of particles moving round a storage ring. Particles in a storage ring also perform synchrotron oscillations, which are oscillations longitudinally (i.e. parallel to the reference trajectory) relative to the reference particle. The number of complete betatron oscillations made by a particle in one revolution of a storage ring (known as the *betatron tune*) is typically quite large—the betatron wavelength may be only a few metres, so if the circumference of the ring is around 100 m, the betatron tunes (horizontal and vertical) may be 10 or more. By contrast, a particle may take many turns of a storage ring to complete a single synchrotron oscillation, i.e. the *synchrotron tune* may be of order 0.01, or less. For that reason, and also because the mechanism behind the oscillations is rather different, synchrotron oscillations are often treated separately from betatron oscillations.

Betatron oscillations result from the focusing effects of quadrupole magnets; synchrotron oscillations, on the other hand, result from the effects of RF cavities. To understand the mechanism of synchrotron oscillations, consider a particle moving around a storage ring slightly ahead of the reference particle. For simplicity, suppose that the storage ring contains a single RF cavity. By definition, the reference particle arrives at the RF cavity at a phase of the oscillation of the cavity voltage such that it receives exactly the right amount of energy from the cavity to replace the energy lost through synchrotron radiation. However, the particle that is ahead of the reference particle arrives too early; depending on whether the cavity voltage is increasing or decreasing, the 'early' particle will receive too little or too much energy to replace that lost by synchrotron radiation. Suppose that this particle receives too much energy. Then, over successive turns, its energy will continue to increase, with the result that (assuming the phase slip factor is positive) its revolution period will also increase. Eventually, this particle will fall behind the reference particle. When that occurs, the particle will now gain less energy from the RF cavity than it needs to replace the energy lost by synchrotron radiation, and its revolution period will start to decrease until, once again, it moves ahead of the reference particle. At that point, the cycle repeats. The regular, periodic motion of a particle moving in front of, and then behind the reference particle is known as a synchrotron oscillation. The process is illustrated in figure 2.8.

The stability of synchrotron oscillations depends on whether the reference particle arrives at the RF cavity when the voltage is increasing or decreasing. In the description given above, we assumed that a particle ahead of the reference particle (i.e. arriving too early at the RF cavity) would receive more energy than the reference particle, in other words that the voltage was decreasing. If the voltage is increasing, on the other hand, a particle arriving too early at the cavity would receive too little energy. If the phase slip factor is positive (which is usual in electron storage rings) then this means the revolution period decreases, so the particle moves further and further ahead of the reference particle. In this situation, there will not be any stable synchrotron oscillations. For a ring above the transition energy (so that the phase slip factor is positive), stable motion in an electron storage ring depends on

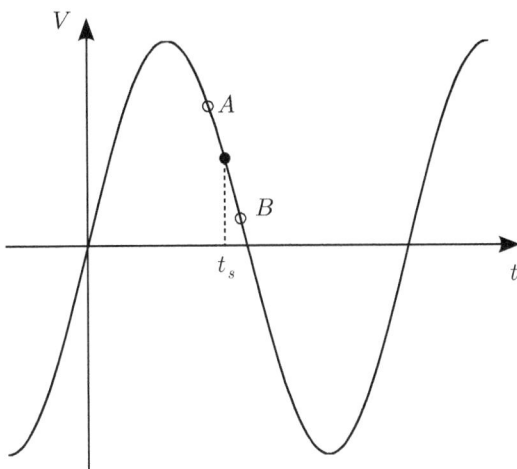

Figure 2.8. Phase stability in a synchrotron storage ring. The voltage in an RF cavity varies sinusoidally—a particle arriving at the cavity at the synchronous phase (i.e. at time t_s) receives exactly the right amount of energy from the cavity to replace the energy lost by synchrotron radiation. A particle arriving early (at point A) on the *falling* slope of the RF voltage receives too much energy: in a storage ring above transition, this leads to an increase in the revolution period, and eventually the particle falls back towards the synchronous phase. Conversely, a particle arriving late (at point B) on the falling slope of the RF receives too little energy, leading to a decrease in the revolution period, and again the particle moves towards the synchronous phase. Thus in a storage ring above transition, particles perform stable longitudinal oscillations about a point (the synchronous phase) on the falling slope of the RF voltage.

particles arriving at an RF cavity when the voltage in the cavity is decreasing; below transition energy, particles must arrive at a cavity when the voltage is increasing. The existence of stable synchrotron oscillations resulting from the mechanism described above is known as *phase stability*.

The fact that the voltage in an RF cavity varies sinusoidally in time implies that only particles within a certain distance of the reference particle can perform stable synchrotron oscillations—particles beyond this distance (sometimes said to be 'outside the RF bucket') will be lost from the beam. As a result, particles in an electron storage ring are grouped in bunches, with each bunch having a certain length determined by (amongst other factors) the RF voltage and phase, and the momentum compaction factor. The maximum number of bunches in a storage ring is equal to the revolution period divided by the RF period, i.e. the number of oscillations of the cavity voltage in the time taken for the reference particle to make a complete circuit of the ring. This is known as the *harmonic number, h*:

$$h = \frac{T_0}{T_{RF}}, \tag{2.79}$$

where T_0 is the revolution period for the reference particle, and T_{RF} is the period of the voltage oscillation in the RF cavity. The harmonic number is necessarily an integer. In an electron storage ring, if a (small) change is made to the RF frequency so that the harmonic number is no longer an exact integer, then the beam (through synchrotron radiation effects) makes a corresponding change in energy, so that the revolution period changes, and the harmonic number is once again an exact integer. The synchronisation between the beam revolution period and the period of voltage oscillations in the RF cavities is a fundamental principle of the operation of synchrotron storage rings.

It is worth mentioning that in addition to oscillations in time, particles make oscillations in energy as they perform synchrotron motion in a storage ring. If the energy were considered more carefully in the above discussion of synchrotron oscillations we would conclude that the energy oscillations are out of phase (by 90°) with the oscillations in time. This means that if we plot the energy deviation δ against the longitudinal co-ordinate z of a particle on successive turns, we will eventually map out an ellipse in longitudinal phase space. The number of synchrotron oscillations completed in one revolution around the ring is the synchrotron tune; as already mentioned, this is typically a small number, much less than one. A synchrotron tune of 0.01 would mean that a particle has to make 100 turns of the ring before it completes a full synchrotron oscillation. A detailed analysis of synchrotron motion leads to the following formula for the angular frequency ω_s of synchrotron oscillations:

$$\omega_s = \frac{c}{C_0} \sqrt{-\frac{qV_{RF}}{cP_0} \frac{\omega_{RF}C_0}{c} \eta_p \cos(\phi_s)}, \qquad (2.80)$$

where q is the electric charge on a particle in the beam, the voltage in the RF cavity has amplitude V_{RF} and angular frequency ω_{RF}, P_0 is the momentum of the reference particle, C_0 is the circumference of the closed orbit for the reference particle, η_p is the phase slip factor, and ϕ_s is the synchronous phase (the phase at which a particle arriving at the RF cavity receives exactly the amount of energy needed to replace the energy lost by synchrotron radiation). The voltage in the cavity varies as

$$V = V_{RF} \sin\left(\phi_s - \frac{\omega_{RF}z}{c}\right). \qquad (2.81)$$

The synchrotron tune is the number of synchrotron oscillations completed per turn of the ring:

$$\nu_s = \frac{c}{C_0} \frac{\omega_s}{2\pi}. \qquad (2.82)$$

Notice that for stability of synchrotron oscillations in a storage ring above transition energy ($\eta_p > 0$), we require $\cos(\phi_s) < 0$, which implies that $\pi/2 < \phi_s < \pi$, i.e. that the voltage in the cavity is falling when the reference particle arrives at the cavity, as expected.

References

[1] Wiedemann H 2015 *Particle Accelerator Physics* 4th edn (New York: Springer)
[2] Conte M and Mackay W W 2008 *An Introduction to the Physics of Particle Accelerators* 2nd edn (Singapore: World Scientific)

[3] Lee S Y 2011 *Accelerator Physics* 3rd edn (Singapore: World Scientific)

[4] Wolski A 2014 *Beam Dynamics in High Energy Particle Accelerators* (London, UK: Imperial College Press)

[5] Bernal S 2016 *A Practical Introduction to Beam Physics and Particle Accelerators* (San Rafael, California, USA: Morgan & Claypool Publishers) (Bristol, UK: IOP Concise Physics, IOP Publishing)

[6] Wolski Andrzej 2011 Theory of electromagnetic fields *Proc. CERN Accelerator School 2010: RF for accelerators (Ebeltoft, Denmark, 8–17 June 2010)*, ed Roger Bailey, CERN–2011–007 (Geneva, Switzerland: CERN) pp 15–66

[7] Goldstein H, Poole C P Jr and Safko J L 2001 *Classical Mechanics* 3rd edn (Boston, MA, USA: Addison-Wesley)

[8] Chambers E E 1968 Particle motion in a standing wave linear accelerator *Proc. 1968 Summer Study on Superconducting Devices and Accelerators (Brookhaven, NY, USA)* pp79–100

[9] Rosenzweig J and Serafini L 2012 Transverse particle motion in radio-frequency linear accelerators *Phys. Rev.* E **49** 1599–602

[10] Muratori B D, Jones J K and Wolski A 2015 Analytical expressions for fringe fields in multipole magnets *Phys. Rev. Spec. Top. - Accel. Beams* **18** 064001

[11] Hill G W 1886 On the part of the motion of lunar perigee which is a function of the mean motions of the Sun and the Moon *Acta Math.* **8** 1–36

[12] Magnus W and Winkler S 2004 *Hill's Equation* (Mineola, New York, USA: Dover) (Reprinted from the original text published by Wiley, New York, 1966)

[13] Twiss R Q and Frank N H 1949 Orbital stability in a proton synchrotron *Rev. Sci. Instrum.* **20** 1

[14] Courant Ernest D and Snyder H S 1958 Theory of the alternating-gradient synchrotron *Ann. Phys.* **3** 1–48

[15] Edwards D A and Teng L C 1973 Parameterization of linear coupled motion in periodic systems *IEEE Trans. Nucl. Sci.* **20** 885–8

[16] Sagan D and Rubin D 1999 Linear analysis of coupled lattices *Phys. Rev. Spec. Top. - Accel. Beams* **2** 074001

[17] Wolski A 2006 Alternative approach to general coupled linear optics *Phys. Rev. Spec. Top. - Accel. Beams* **9** 024001

[18] Dowd R, Boland M, LeBlanc G and Tan Y-R E 2011 Achievement of ultralow emittance coupling in the Australian Synchrotron storage ring *Phys. Rev. Spec. Top. - Accel. Beams* **14** 012804

[19] Minty M G and Zimmermann F 2003 *Measurement and Control of Charged Particle Beams* (Berlin: Springer)

IOP Concise Physics

Introduction to Beam Dynamics in High-Energy Electron Storage Rings

Andrzej Wolski

Chapter 3

Synchrotron radiation

3.1 Features of synchrotron radiation

When a charged particle is accelerated, it produces electromagnetic radiation. This can be understood in terms of the fields around a charged particle. For example, consider an electron at rest—the electric field around the electron is static, and there is no radiation. However, if the electron is oscillating around a fixed point, then the motion of the electron leads to a regular variation of the electric field. There will also now be a magnetic field, since a moving charge constitutes a current. The periodic variations in the electric and magnetic fields will propagate away from the electron at the speed of light, carrying energy that may be observed as electromagnetic radiation.

Any acceleration of a charged particle leads to disturbances in the fields around the particle, and hence to radiation. If the particle is moving relativistically, the radiation is known as *synchrotron radiation*. In an electron storage ring, particles accelerate (by changing their direction of motion, though not their speed) as they move through the magnetic fields in the ring. Synchrotron radiation is then produced, for example, from the particles moving through the dipole magnets that steer the beam around the ring. The amount of radiation produced depends on the strengths of the field seen by the beam; in quadrupole (and higher-order multipole) magnets, the fields seen by particles in the beam are generally much weaker than the fields seen by particles in the dipole magnets, so the amount of synchrotron radiation produced in these magnets is (except in some special cases) negligible. Similarly, although synchrotron radiation is produced by particles in a linac, at ultrarelativistic velocities the rate of acceleration is small, so the amount of radiation is also small.

The amount of synchrotron radiation produced by accelerating a charged particle depends on the charge-to-mass ratio of the particle as well as on the rate of

doi:10.1088/978-1-6817-4989-1ch3   3-1

acceleration. Although synchrotron radiation is produced by protons in storage rings, since protons have much larger mass than electrons, the amount of radiation from protons in a storage ring is small, and can usually be ignored.

In this chapter, we shall be mainly concerned with the impact of synchrotron radiation on particles producing that radiation in an electron storage ring. As we shall see in the following sections, one of the main effects of synchrotron radiation is to lead to the beam reaching equilibrium values for the beam emittances, determined by the beam energy and the arrangement of magnets in the storage ring. We shall not be primarily concerned with the properties of the radiation itself; however, since the properties of synchrotron radiation have proved extremely valuable for science and engineering in a wide range of fields [1–4], it is worth mentioning at this point some of the features of synchrotron radiation from electron storage rings. Many of the important properties of synchrotron radiation were derived by Schwinger [5]. Further information can be found in a number of references, including (for example) [6–9].

3.1.1 Radiation power spectrum

The radiation power from an accelerated charged particle can be found from classical electromagnetic theory. A particle with energy E following a curved trajectory with radius ρ emits radiation with power [10]

$$P_\gamma = \frac{C_\gamma}{2\pi} \frac{cE^4}{\rho^2}. \tag{3.1}$$

C_γ is a constant given by

$$C_\gamma = \frac{q^2}{4\epsilon_0 (mc^2)^4}, \tag{3.2}$$

where q is the electric charge of the particle, m is the mass of the particle, and ϵ_0 is the permittivity of free space. For electrons, $C_\gamma \approx 8.846 \times 10^{-5}$ m GeV^{-3}. The total radiation power (in kilowatts) from the dipole magnets in an electron storage ring is then found to be

$$P_{\text{dipole}} \, [\text{kW}] \approx 88.46 \frac{(E \, [\text{GeV}])^4 I \, [\text{A}]}{\rho \, [\text{m}]}, \tag{3.3}$$

where I is the electron beam current. For typical parameters in a storage ring for a third-generation synchrotron light source, we may consider a beam energy of 3 GeV, a dipole bending radius of 6 m, and a beam current of 0.3 A—the radiation power from the dipole magnets in this case would be approximately 360 kW. Much of this power will be in the ultra-violet or soft x-ray region of the electromagnetic spectrum. Electron storage rings can provide much larger amounts of power in this part of the spectrum than can be produced using non-accelerator based short-wavelength radiation sources.

The synchrotron radiation from a charged particle in a dipole magnet extends over a broad range of wavelengths. In synchrotron light sources, there are typically significant amounts of power in the part of the electromagnetic spectrum ranging from the infra-red up to the ultra-violet or soft x-ray regions. The spectrum can be calculated from the electric and magnetic fields produced by a charged particle moving through a dipole magnet, where the fields are observed at a fixed point some distance downstream from the magnet. It is found that the power spectrum (power dP emitted within a small frequency range $d\omega$) is described by the formula

$$\frac{dP}{d\omega} = \frac{\sqrt{3}\,e^2}{4\pi\epsilon_0 c}\gamma\frac{\omega}{\omega_c}\int_{\omega/\omega_c}^{\infty} K_{5/3}(x)\,dx, \tag{3.4}$$

where $K_{5/3}(x)$ is a modified Bessel function, ω is the angular frequency of the radiation, and ω_c is the *critical frequency* given by

$$\omega_c = \frac{3c}{2\rho}\gamma^3. \tag{3.5}$$

The radiation power spectrum for a single electron moving through a magnetic field is shown in figure 3.1: note that the strength of the field is accounted for in the scale of the horizontal axis, which gives the radiation frequency ω divided by the critical frequency ω_c. The peak in the radiation spectrum occurs close to the critical frequency. For an electron with energy 3 GeV moving through a dipole with bending radius 6 m (field strength approximately 1.67 T), the radiation wavelength at the critical frequency is roughly 0.12 nm, which is in the soft x-ray region of the electromagnetic spectrum. Above the critical frequency, the radiation power falls

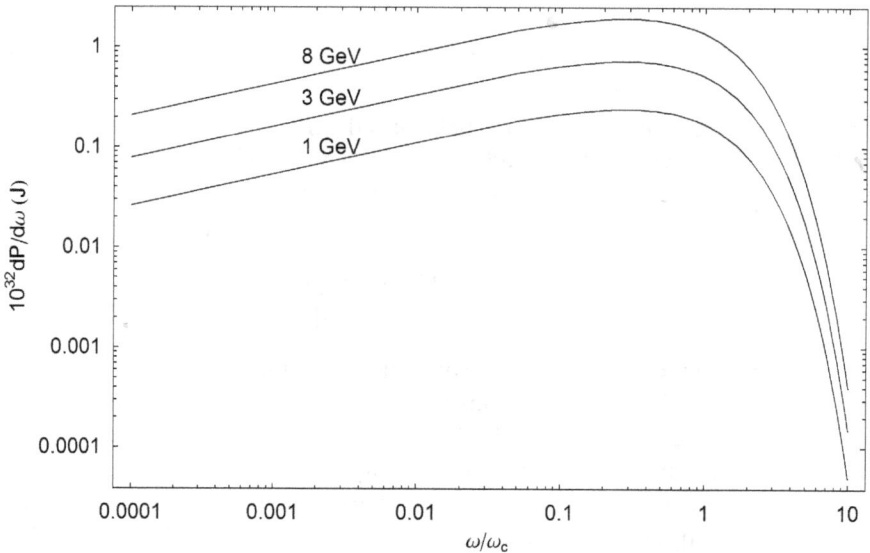

Figure 3.1. Synchrotron radiation power spectrum for a single electron moving through a magnetic field. The curves correspond to electron energies of 1 GeV, 3 GeV, and 8 GeV. The radiation frequency is ω, and the critical frequency is ω_c, given by equation (3.5). The spectral power peaks close to the critical frequency; note that for fixed bending radius, the critical frequency increases as the third power of the electron energy.

rapidly with increasing frequency. For efficient production of hard x-ray radiation, which is important for a number of scientific fields, insertion devices (undulators and wigglers [9]) can be used. These devices generally consist of a sequence of short dipoles (length of order of a few centimetres) of alternating polarity, that cause particles to follow a sinusoidal trajectory as they pass along the device.

As we discuss later (in section 3.1.3) the synchrotron radiation from a single particle is emitted in a narrow cone around the instantaneous direction of motion of the particle. The opening angle of the radiation is roughly $1/\gamma$, where γ is the relativistic factor for the particle. In a dipole magnet, particles emit radiation along the entire length of the magnet, and since the bending angle of a dipole in a storage ring is typically large compared to $1/\gamma$, the radiation from the beam in a dipole takes the form of a 'fan' with narrow vertical divergence, but horizontal divergence roughly equal to the dipole bending angle.

Although synchrotron radiation from dipoles is often very useful for many light source users, third-generation light sources are designed to optimise the radiation from insertion devices. Wigglers are designed so that the amplitude of the trajectory oscillation around the axis of the device is large. The radiation power spectrum from a wiggler is then similar to that from a dipole magnet, and the horizontal divergence is dominated by the change in angle of the beam trajectory along the device. In an undulator, however, the amplitude of the trajectory oscillation around the axis is much smaller. This leads to interference effects between the radiation produced in successive periods of the array of magnets, and results in a radiation power spectrum dominated by a series of sharp spikes (essentially, a line spectrum). The angular divergence of undulator radiation is also dominated by the intrinsic opening angle of radiation from individual particles (of order $1/\gamma$) rather than by the trajectory of the beam.

A detailed analysis of radiation emission in an undulator leads to the *undulator equation* giving the wavelengths λ_n of the spectral lines,

$$\lambda_n = \frac{\lambda_u}{2n\gamma^2}\left(1 + \frac{K^2}{2} + \theta^2\gamma^2\right), \tag{3.6}$$

where λ_u is the undulator period (the distance over which the field in the undulator repeats), γ is the relativistic factor for the electron beam, θ is the angle with respect to the axis of the undulator at which the radiation is observed, and n is an integer corresponding to the different harmonics (lines) in the radiation spectrum. K is the deflection parameter that characterises the amplitude of the oscillation of the trajectory around the axis of the insertion device: $K < 1$ for an undulator and $K > 1$ for a wiggler (see section 3.1.3).

3.1.2 Brightness

Although the radiation power is a significant quantity for many applications of synchrotron radiation, a more important figure of merit is often the radiation beam *brightness*. The brightness is defined as the amount of radiation in a given frequency range, per unit area in phase space of the source of the radiation. The area of phase space occupied by electrons in a storage ring is quantified by the emittance (see section 2.3), which is a constant as the beam moves around the ring: the photon beam brightness is therefore independent of the location of the source of the radiation in the storage ring. Since the radiation from each electron is emitted in a narrow cone around the instantaneous direction of motion of the electron, the brightness is also a measure of the power per unit phase space area of the radiation beam, and is a constant as the beam is transported along a synchrotron radiation beam line from the storage ring to an experimental area.

The brightness can be calculated from the formula:

$$B = \frac{\Phi_\gamma}{4\pi^2 \Sigma_x \Sigma_{x'} \Sigma_y \Sigma_{y'} (d\omega/\omega)},$$ (3.7)

where Φ_γ is the photon flux (number of photons produced per second) at angular frequency ω, Σ_x and $\Sigma_{x'}$ are given (respectively) by summing in quadrature the electron and photon beam size, and the electron and photon beam divergence:

$$\Sigma_x = \sqrt{\sigma_x^2 + \sigma_r^2}, \quad \text{and} \quad \Sigma_{x'} = \sqrt{\sigma_{x'}^2 + \sigma_{r'}^2},$$ (3.8)

and similarly for Σ_y and $\Sigma_{y'}$. Here, we assume that the electron beam has a Gaussian distribution, so that $\sigma_x^2 = \langle x^2 \rangle$, and $\sigma_{x'}^2 = \langle x'^2 \rangle$. Under ideal conditions, the photon beam size and divergence in an undulator are limited by diffraction effects, in which case they are given by

$$\sigma_r = \frac{1}{2\pi}\sqrt{\frac{\lambda L}{2}}, \quad \text{and} \quad \sigma_{r'} = \sqrt{\frac{\lambda}{2L}},$$ (3.9)

where λ is the wavelength of the radiation, and L is the length of the undulator.

3.1.3 Opening angle of the radiation beam

One of the important features of synchrotron radiation for many applications is the intensity of the radiation—it is often helpful in experiments to have a beam with a small divergence. Radiation from a relativistic charged particle is emitted with a spatial distribution characterised by an opening angle (divergence) of $1/\gamma$ around the instantaneous direction of motion of the particle. In a third-generation synchrotron light source, a typical electron energy is 3 GeV, so the relativistic factor γ is of order 6000; the synchrotron radiation can then have an opening angle of a fraction of a milliradian, i.e. of order 0.01°. However, in a dipole magnet particles radiate continually as they move through the magnet; although the vertical divergence of the radiation beam may be small, in the horizontal (bending) plane, the opening angle of the radiation essentially becomes equal to the bending angle of the magnet, which may be much larger, of order 10°. Although it is possible to reduce the horizontal divergence by collimating the beam, so that the radiation is effectively observed from only a small section of the beam trajectory through the dipole, this means a significant loss of intensity.

As an alternative to dipole magnets, it is possible to use magnets with field shapes specifically designed to generate high-intensity radiation. Such magnets are generally known as *insertion devices*, and are classed as either *undulators* or *wigglers* [9]. In either case, the basic structure is a sequence of short dipole magnets of alternating polarity. With an appropriate design, the magnet causes no overall deflection of the beam, though each particle follows an oscillating trajectory as it travels through the insertion device. The radiation from particles within each period adds to the overall intensity of the radiation beam. The overall radiation power can be similar to (or greater than) that produced by a dipole magnet, but because the beam undergoes only a small deflection within each period of an insertion device, the divergence of the radiation beam from an insertion device will be much smaller than the divergence from a dipole. In an undulator, the parameters are such that the divergence of the radiation beam is dominated by the opening angle of the radiation from a single particle, i.e. $1/\gamma$. This generally means that the period of the insertion device (the length of a single unit consisting of a pair of dipoles of opposite polarity) is short. In a wiggler, the period is long enough (and the field strong enough) that the electron beam trajectory undergoes significant deflection within the body of the insertion device, so the divergence of the radiation beam is dominated by maximum angle between the electron beam trajectory and a straight line down the centre of the insertion device.

Consider an insertion device that produces a magnetic field varying sinusoidally along the length of the device, with amplitude B_0 and period λ_u. Electrons will follow a sinusoidal trajectory through the device; solving the equations of motion shows that the peak angular deflection will be K/γ, where γ is the relativistic factor, and K is the *deflection parameter* of the insertion device, given by

$$K = \frac{eB_0}{mc} \frac{\lambda_u}{2\pi} \approx 93.36 B_0 \,[\text{T}]\, \lambda_u \,[\text{m}]. \tag{3.10}$$

In an undulator, $K < 1$; the divergence of the radiation beam will then be dominated by the intrinsic opening angle of the radiation ($1/\gamma$) from an individual particle. In a wiggler, however, $K > 1$ and the divergence of the radiation beam will be dominated by the angular deflection of particle trajectories through the wiggler. For a peak field of 0.2 T, an undulator will have a period less than roughly 5 cm.

3.1.4 Polarisation

Another feature of synchrotron radiation that can be important for some applications is its polarisation. Synchrotron radiation from a dipole magnet bending a beam in the horizontal plane is horizontally polarised (i.e. the electric field oscillates in the horizontal plane) when observed in the horizontal plane. Radiation emitted at (small) angles above or below the horizontal plane has a vertical component of polarisation in addition to the horizontal component.

Insertion devices in which the field polarity alternates along the length of the device produce radiation with similar polarisation properties to dipoles. This is not surprising, since such insertion devices consist essentially of sequences of dipole

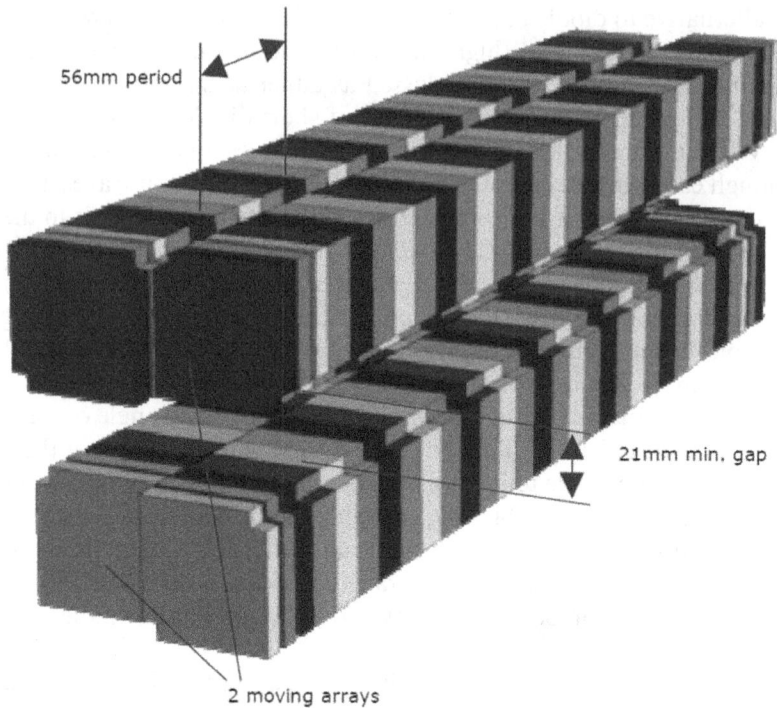

Figure 3.2. Schematic of the arrangement of magnetic material in an APPLE-II type undulator [11]. The different colour blocks represent permanent magnetic material with different orientations of the magnetisation. By adjusting the relative positions of the arrays longitudinally, it is possible to control the magnetic field along the beam path so that electrons follow a planar or helical periodic trajectory. This leads to the production of synchrotron radiation with planar or circular polarisation.

magnets. However, because insertion devices are generally constructed so that there is no overall deflection of the electron beam, the magnetic field in an insertion device does not have to be purely vertical; the orientation of the polarisation from an insertion device could, in principle, be at any angle.

More sophisticated insertion devices allow the field to be controlled to produce elliptical polarisation of the synchrotron radiation, in addition to planar polarisation. This can be achieved by moving the magnetic poles in the insertion device relative to one another longitudinally (see figure 3.2). The direction of the magnetic field can then be made to rotate with distance along the insertion device. Electrons follow a helical trajectory as they travel through the insertion device, producing radiation with an elliptical polarisation, which can be useful for some applications.

3.2 Radiation damping and quantum excitation

In this section, we shall discuss how the effects of synchrotron radiation lead to the horizontal, vertical, and longitudinal emittances of a beam in an electron storage ring reaching equilibrium values determined by the beam energy and the design of the magnetic lattice. Briefly, the energy loss from synchrotron radiation leads to an exponential decrease (damping) of the amplitudes of synchrotron and betatron oscillations of any electron in a storage ring. However, the quantum nature of the radiation prevents the amplitudes damping to zero: the random emission of photons results in some excitation of synchrotron and betatron oscillations. The equilibrium emittances are determined by the balance between radiation damping and quantum excitation. We begin by discussing the damping effects of radiation, and then we consider the quantum excitation. The different physical processes involved in synchrotron and betatron oscillations make it convenient to consider longitudinal and transverse motion separately.

3.2.1 Damping of synchrotron oscillations

In section 2.8 we saw that the combined effects of the RF cavities in a storage ring and phase slip (from dispersion in the dipole magnets) led to particles performing synchrotron oscillations. In that discussion, we neglected the effects of synchrotron radiation. We shall now include radiation effects, and consider their effect on the synchrotron oscillations performed by particles as they move around a storage ring.

First of all, let us write down the equations of motion for the longitudinal co-ordinate z and the energy deviation δ of an electron in a storage ring. On each turn, the electron acquires energy from the RF cavities, and loses energy through synchrotron radiation. Averaging the energy gain and loss over a single turn, the rate of change of the energy deviation can be written

$$\frac{d\delta}{dt} = \frac{qV_{\text{RF}}}{E_0 T_0} \sin\left(\phi_{\text{s}} - \frac{\omega_{\text{RF}} z}{c}\right) - \frac{U}{E_0 T_0}, \tag{3.11}$$

where q is the electric charge of the particle, V_{RF} and ω_{RF} are the amplitude and angular frequency (respectively) of the voltage in the cavity, E_0 is the reference energy, T_0 is the revolution period, and U is the total energy lost through synchrotron radiation. The synchronous phase ϕ_{s} is defined so that the energy gained by the reference particle (which has $z = 0$) in an RF cavity is exactly matched by the energy U lost through synchrotron radiation over one turn.

The rate of change of the longitudinal co-ordinate is related to the energy deviation by the phase slip factor η_{p} (2.77). Again averaging over a single turn, the rate of change of the longitudinal co-ordinate is

$$\frac{dz}{dt} = -\eta_{\text{p}} c \delta. \tag{3.12}$$

The above equations, (3.11) and (3.12), are the longitudinal equations of motion for a particle in a synchrotron storage ring, including the effects of synchrotron radiation. If we assume that z is small, so that $\omega_{\text{RF}} z/c \ll 1$, we can make the approximation

$$\sin\left(\phi_{\text{s}} - \frac{\omega_{\text{RF}} z}{c}\right) \approx \sin(\phi_{\text{s}}) - \cos(\phi_{\text{s}})\frac{\omega_{\text{RF}}}{c} z. \tag{3.13}$$

The equations of motion are then linear in the variables z and δ, and can be combined to give the following equation for the energy deviation:

$$\frac{d^2\delta}{dt^2} + 2\alpha_E \frac{d\delta}{dt} + \omega_s^2 \delta = 0. \tag{3.14}$$

The synchrotron frequency is $\omega_s = 2\pi\nu_s/T_0$, with ν_s the synchrotron tune (2.82). The constant α_E is given by the change of energy lost through synchrotron radiation with respect to a change in the energy of the particle, evaluated at the reference energy:

$$\alpha_E = \frac{1}{2T_0}\frac{dU}{dE}\bigg|_{E=E_0}. \tag{3.15}$$

The equation of motion for the energy deviation (3.14) is the equation for a damped harmonic oscillator, with angular frequency ω_s and damping time $\tau_z = 1/\alpha_E$. The solution to the equation of motion can be written

$$\delta(t) = \delta_0 e^{-\alpha_E t}\sin(\omega_s t + \theta_0), \tag{3.16}$$

where δ_0 is the initial amplitude (the amplitude of the energy oscillations at time $t = 0$), and θ_0 is the initial phase of the oscillation. We find that the longitudinal co-ordinate z obeys a similar equation of motion to the energy deviation, and that:

$$z(t) = \frac{\eta_p c}{\omega_s}\delta_0 e^{-\alpha_E t}\sin(\omega_s t + \theta_0). \tag{3.17}$$

The main significance of these results is that synchrotron radiation leads to damping (exponential decay of the amplitude) of synchrotron oscillations. To complete the description of the motion, however, we need to find an expression for the damping constant α_E in terms of the parameters of the storage ring and the electron beam. This can be accomplished using the expression we have already quoted (3.1) for the synchrotron radiation power. The energy loss per turn is found by integrating the radiation power over a complete turn of the ring, taking into account the dependence of the revolution period on the energy deviation (because of the presence of dispersion η_x). The result of the calculation is that for a particle with zero energy deviation, the energy loss per turn is

$$U_0 = \frac{C_\gamma}{2\pi}E_0^4 I_2, \tag{3.18}$$

where I_2 is the *second synchrotron radiation integral*, obtained by integrating the radius of curvature of the electron trajectory over the circumference of the ring

$$I_2 = \int_0^{C_0}\frac{1}{\rho^2}\,ds. \tag{3.19}$$

The damping constant α_E depends on the derivative of the energy loss per turn with respect to energy; this can be found from the same calculation as that for U_0, but not imposing the condition $\delta = 0$. The result is that the *longitudinal damping time* τ_z is given by

$$\tau_z = \frac{1}{\alpha_E} = \frac{2}{j_z} \frac{E_0}{U_0} T_0, \tag{3.20}$$

where j_z is the *longitudinal damping partition number*:

$$j_z = 2 + \frac{I_4}{I_2}. \tag{3.21}$$

I_4 is the *fourth synchrotron radiation integral*:

$$I_4 = \int_0^{C_0} \frac{\eta_x}{\rho} \left(\frac{1}{\rho^2} + 2k_1 \right) ds, \tag{3.22}$$

where $k_1 = (q/P_0) \, \partial B_y/\partial x$ is the local quadrupole focusing strength (2.11). Note that the focusing strength only affects the value of I_4 if the quadrupole field component appears in a dipole magnet: this takes into account the fact that in this case, the field strength will vary with the energy of the particle, because of dispersion in the dipole magnet. If the dipole magnets in a storage ring have no quadrupole component (i.e. the field is uniform, and does not vary with x or y), then $I_4 \ll I_2$, and $j_z \approx 2$.

3.2.2 Damping of betatron oscillations

In our analysis of the effects of synchrotron radiation on synchrotron motion in a storage ring, we were able to average the radiation energy loss over a complete revolution of a particle around the ring: this was a valid approach, because in most cases the synchrotron tune is small, $\nu_s \ll 1$, so that the changes in the co-ordinate z and energy deviation δ are small over one turn. In the case of betatron oscillations, however, particles will usually complete many oscillations over a single turn of the ring, so we cannot assume that we can average the effects of synchrotron radiation over one turn. To understand the impact of synchrotron radiation on betatron motion, we need to take another approach. We shall first consider the vertical betatron motion of particles in a storage ring: this is simpler than the horizontal betatron motion, since for the latter case we need to take into account effects from dispersion.

In section 2.2.4 we defined the betatron action as a measure of the amplitude of betatron oscillations: the action is defined in such a way that it remains constant for a given particle, as the particle moves around the ring. However, in the discussion in section 2.2.4, we ignored any effects from synchrotron radiation. Suppose that a particle emits some synchrotron radiation as it passes through a dipole magnet. The energy and momentum of the particle will fall as a result. Since the radiation is emitted (for ultrarelativistic particles, with $\gamma \gg 1$) along the instantaneous direction of motion of the particle, the emission of synchrotron radiation will not change the direction in which the particle is moving. This means that each component of the momentum will scale by the same factor. If the change in momentum is Δp and the particle initially has a momentum close to the reference momentum P_0, then each component of the momentum vector will scale by a factor (approximately) $1 - \Delta p/P_0$. In particular, the vertical momentum p_y will undergo a change $\Delta p_y = -\Delta p/P_0$. This will lead to a change in the vertical betatron action J_y for the particle; the change depends on the betatron phase, but averaging over all particles in the beam (which we assume to be uniformly distributed in betatron phase) we find that the average change in the vertical betatron action is

$$\langle \Delta J_y \rangle = -\varepsilon_y \frac{\Delta p}{P_0}, \tag{3.23}$$

where $\varepsilon_y = \langle J_y \rangle$ is the vertical emittance of the beam.

If the total change in betatron action over a single turn of the ring is small, we can find the rate of change of the betatron action by averaging the changes over one revolution: we were not able to do this with the co-ordinate y and momentum p_y directly, because particles will usually complete many betatron oscillations in one turn, so the changes in y and p_y can be large over a single revolution period. However, assuming that the change in the betatron action over one turn from synchrotron radiation for any particle is small, then averaging over one turn is a valid approach. Integrating the momentum loss around the ring and dividing by the revolution period T_0 gives

$$\frac{d\varepsilon_y}{dt} = -\frac{\varepsilon_y}{T_0} \int_0^{C_0} \frac{dp}{P_0} \approx -\frac{U_0}{E_0 T_0}\varepsilon_y = -\frac{2}{\tau_y}\varepsilon_y, \tag{3.24}$$

where we have assumed that the particle is ultra-relativistic, so that the energy $E_0 \approx P_0 c$, and we define the *vertical damping time*

$$\tau_y = 2\frac{E_0}{U_0}T_0. \tag{3.25}$$

The vertical emittance falls exponentially, with damping time $\tau_y/2$:

$$\varepsilon_y(t) = \varepsilon_y(0)e^{-2t/\tau_y}. \tag{3.26}$$

Since the vertical emittance is a measure of the vertical betatron amplitudes of particles in the beam, we see that the impact of synchrotron radiation on the vertical betatron motion is to cause an exponential decay in the amplitude of the motion.

The energy lost by particles through synchrotron radiation is replaced in electron storage rings by RF cavities (see section 1.2.2). Since the gain in energy by a particle in an RF cavity is accompanied by an increase in momentum, we should consider the impact of the change in momentum on the betatron action of the particle (and hence on the emittance of the beam). However, RF cavities are generally designed so that the accelerating field is parallel to the reference trajectory. In that case, although the momentum of a particle may increase as it passes through the cavity, only the longitudinal component will change, and the transverse components (p_x and p_y) will remain the same. The horizontal and vertical betatron actions of a particle passing through an RF cavity will therefore be unchanged by the fields in the cavity, and the cavity will have no effect on the horizontal or vertical emittance.

Fundamentally, synchrotron radiation affects the horizontal emittance in the same way as the vertical emittance: when a particle emits synchrotron radiation, there will be a reduction in its horizontal momentum, and a change in the horizontal betatron action of the particle. Averaged over all the particles in the beam, this leads to an exponential reduction in the horizontal emittance. However, calculating the horizontal damping time is more complicated than calculating the vertical damping time, for three reasons. First, the horizontal motion is strongly coupled to the longitudinal motion through the dispersion. This means that as a particle emits electromagnetic radiation, there will be a change in the closed orbit about which the trajectory of the particle oscillates. This results in a change in the betatron amplitude, in addition to the change resulting directly from the change in horizontal momentum of the particle. Second, the curvature (in the horizontal plane) of the reference trajectory means that the length of the closed orbit depends on the horizontal co-ordinate of a particle, and hence on the horizontal betatron action. Finally, dipole magnets are sometimes constructed with a quadrupole field component, so the field strength seen by a particle in a dipole magnet depends on the horizontal co-ordinate of the particle, and therefore on the horizontal betatron action.

Taking the relevant effects into account, we nonetheless find that the horizontal emittance damps exponentially, in the same way as the longitudinal and vertical emittances:

$$\varepsilon_x(t) = \varepsilon_x(0)e^{-2t/\tau_x}. \tag{3.27}$$

The horizontal damping time τ_x, however, has to take account of the effects of dispersion, the curvature of the reference trajectory, and the possible presence of a quadrupole field component in the dipole magnets:

$$\tau_x = \frac{2}{j_x}\frac{E_0}{U_0}T_0, \tag{3.28}$$

where j_x is the *horizontal damping partition number*

$$j_x = 1 - \frac{I_4}{I_2}. \tag{3.29}$$

Here, I_2 and I_4 are the second and fourth synchrotron radiation integrals, which we encountered earlier, (3.19) and (3.22).

As we observed in the case of the longitudinal motion, if the dipole magnets have no quadrupole field component, then $I_4 \ll I_2$ and hence $j_x \approx 1$. Inclusion of a quadrupole field component in the dipole magnets, however, makes it possible to 'shift' the radiation damping between the longitudinal and horizontal motion. It is worth noting that $j_x + j_z = 3$, regardless of the design of the dipole magnets. There is, in a sense, only a certain amount of damping provided by synchrotron radiation, and although it is possible (for example) to reduce the horizontal damping time, the longitudinal damping time will increase so that $1/\tau_x + 1/\tau_z$ is constant. This result is known as the *Robinson damping theorem* [12].

3.2.3 Quantum excitation

In our discussion of radiation damping, we assumed that synchrotron radiation could be described by a classical model, and we ignored any effects associated with the fact that the radiation is emitted in quanta (photons). If synchrotron radiation were a purely classical process, then the beam emittances would damp to zero; however, it is observed in practice that the emittances reach non-zero equilibrium values. That this behaviour results from quantum effects is supported by the fact that the expressions for the equilibrium emittances depend on Planck's constant. The emission of photons by electrons in a storage ring excites synchrotron and betatron oscillations. The rate of excitation depends on the rate of emission of photons and on the photon energy distribution, and is not dependent on the amplitude of the synchrotron or betatron oscillations already being performed by an electron. Therefore, the emission of photons does not lead to an exponential increase in the oscillation amplitude, but rather to a linear increase (as a function of time). This means that at some particular value of the emittance, the rate of increase is matched by the rate of damping—the equilibrium emittances are determined by the balance between radiation damping and quantum excitation.

Consider first the effect of photon emission on synchrotron motion. Suppose an electron emits a photon of energy u: this leads to an instantaneous change in the longitudinal co-ordinate z and the energy deviation δ (figure 3.3). The values of z and δ following the photon emission are

$$z' = \frac{\alpha_p c}{\omega_s}\delta_0 \cos(\theta), \quad \text{and} \quad \delta' = \delta_0 \sin(\theta) - \frac{u}{E_0}, \tag{3.30}$$

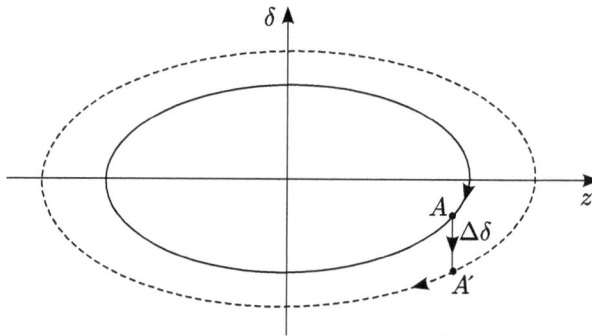

Figure 3.3. A particle performing synchrotron oscillations moves round an ellipse in longitudinal phase space. If the particle emits a photon, the amplitude of the motion (corresponding to the area of the ellipse) can change. In the case shown, the particle is initially moving around the solid ellipse; at point A, it emits a photon of energy u, which leads to a change in the energy deviation δ by an amount $\Delta\delta = u/E_0$, where E_0 is the reference energy. As a result, the particle moves to a point A' in phase space, and now follows a new trajectory, shown by the dashed ellipse, corresponding to synchrotron oscillations with a larger amplitude.

where δ_0 is the initial amplitude of the energy oscillations and θ is the phase in the synchrotron oscillation at which the emission takes place. The change in the amplitude of the energy oscillation can then be found from

$$\delta_0'^2 = \delta_0^2 - 2\delta_0 \frac{u}{E_0} \sin(\theta) + \frac{u^2}{E_0^2}. \tag{3.31}$$

Averaging over all particles in the bunch gives

$$\Delta\sigma_\delta^2 = \frac{\langle u^2 \rangle}{2E_0^2}, \tag{3.32}$$

where $\Delta\sigma_\delta^2 = \langle \delta_0'^2 \rangle - \langle \delta_0^2 \rangle$ is the change in the mean square energy deviation. Suppose that photons in the energy range u to $u + du$ are emitted at a rate $\dot{N}(u)$. Then

$$\frac{d\langle u^2 \rangle}{dt} = \int_0^\infty \dot{N}(u)u^2 du. \tag{3.33}$$

If we include radiation damping as well as the excitation from photon emission, the overall rate of change of the mean square energy deviation is

$$\frac{d\sigma_\delta^2}{dt} = \frac{1}{2E_0^2} \left\langle \int_0^\infty \dot{N}(u)u^2 du \right\rangle_C - \frac{2}{\tau_z}\sigma_\delta^2, \tag{3.34}$$

where $\langle ... \rangle_C$ indicates an average over the circumference of the ring.

The rate of photon emission $\dot{N}(u)$ can be calculated from the synchrotron radiation spectrum (3.4) using the fact that the energy of a single photon of frequency ω is $u = \hbar\omega$. It is found [6] that

$$\int_0^\infty \dot{N}(u)u^2 du = 2C_q\gamma^2 \frac{E}{\rho} P_\gamma, \tag{3.35}$$

where γ is the relativistic factor for a particle with energy E following a trajectory with radius of curvature ρ, P_γ is the synchrotron radiation power (3.1). C_q is the *synchrotron radiation quantum constant*:

$$C_q = \frac{55}{32\sqrt{3}} \frac{\hbar}{mc}, \tag{3.36}$$

where $\hbar$ is Planck's constant divided by 2π. For electrons, $C_q \approx 3.832 \times 10^{-13}$ m. Using the result (3.35) in the expression for the rate of change of the mean square energy deviation (3.34), we find

$$\frac{d\sigma_\delta^2}{dt} = C_q\gamma^2 \frac{2}{j_z\tau_z} \frac{I_3}{I_2} - \frac{2}{\tau_z}\sigma_\delta^2, \tag{3.37}$$

where I_3 is the *third synchrotron radiation integral*:

$$I_3 = \int_0^{C_0} \frac{1}{|\rho|^3} ds. \tag{3.38}$$

The first term in the above expression for the rate of change of the mean square energy deviation (3.37) represents the effects of quantum excitation, and is independent of the energy spread σ_δ—quantum excitation takes place at a constant rate. The second term in (3.37) represents radiation damping, and depends on the energy spread: the larger the energy spread, the faster the damping rate. A balance between quantum excitation and radiation damping is reached when

$$\sigma_\delta^2 = \sigma_{\delta 0}^2 = C_q \gamma^2 \frac{I_3}{j_z I_2}. \tag{3.39}$$

The value of the energy spread $\sigma_{\delta 0}$ given by this expression is the equilibrium value determined by synchrotron radiation effects; this is known as the *natural energy spread*. In practice, the actual energy spread of the electrons in a storage ring may be larger than the natural energy spread, because collective effects (interactions between the electrons) can cause the energy spread to increase. Collective effects will be discussed further in chapter 5.

We have seen that as particles perform energy oscillations in moving round a storage ring, they also perform longitudinal oscillations (i.e. oscillations in the co-ordinate z). From results (3.16) and (3.17) in section 3.2.1, the ratio of the amplitude of the co-ordinate oscillation to the amplitude of the energy oscillation is $\frac{\eta_p c}{\omega_s}$. Therefore, associated with the natural energy spread is the *natural bunch length*:

$$\sigma_{z0} = \frac{\eta_p c}{\omega_s} \sigma_{\delta 0}. \tag{3.40}$$

Quantum excitation of betatron oscillations occurs in a similar way to quantum excitation of synchrotron oscillations: the emission of photons causes changes in the co-ordinates and momenta of particles in the beam, which (on average) leads to a steady increase in the amplitude of betatron oscillations. The process is illustrated in figure 3.4. When we considered radiation damping in section 3.2.2, we gave an expression for the change in the betatron action resulting from emission of radiation carrying momentum Δp (3.23). However, in this expression, we dropped a term in Δp^2; the reason for this is apparent if we write the change in the (horizontal) action in a short time interval Δt:

$$\frac{\langle \Delta J_x \rangle}{\Delta t} = -\frac{\varepsilon_x}{P_0} \frac{\Delta p}{\Delta t} + \frac{w}{P_0^2} \frac{\Delta p^2}{\Delta t}, \tag{3.41}$$

where w is a (complicated) function of the Courant–Snyder parameters and the dispersion. If we take the limit $\Delta t \rightarrow 0$, then the left-hand side of this equation becomes the derivative of the emittance with respect to time, $d\varepsilon_x/dt$, and the ratio $\Delta p / \Delta t$ in the first term on the right-hand side becomes the rate of change of

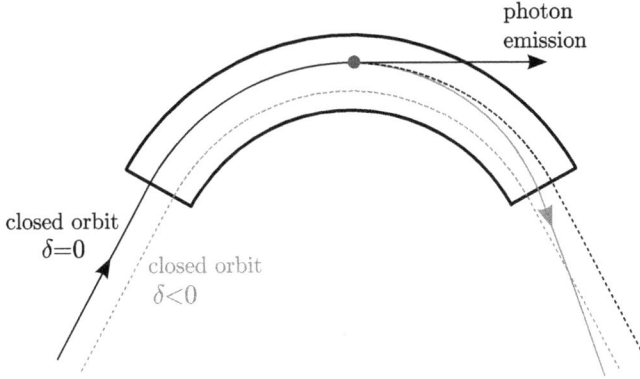

Figure 3.4. Quantum excitation of betatron motion. If a particle with zero energy deviation ($\delta = 0$) is initially following a closed orbit (black line), then, following the emission of a photon, there is a change in the closed orbit, described by the dispersion. Since there is an instantaneous loss of energy, the particle (now with negative energy deviation, $\delta < 0$) is no longer on a closed orbit. As a result of the photon emission, the particle follows a trajectory (solid red line) in which it makes betatron oscillations around the new closed orbit (dashed red line); in other words, there is an increase in the betatron amplitude of the motion of the particle.

momentum, dp/dt. The second term on the right-hand side, however, involves the *square* of a small quantity (the change in momentum) divided by a small quantity (the time interval Δt); in the limit that $\Delta t \rightarrow 0$, we expect this ratio to vanish. However, in the quantum model of radiation, a photon with a finite momentum Δp can be emitted even in a vanishing time interval Δt; if we include quantum effects, therefore, we cannot ignore this term. A full analysis (see, for example, [6]) leads to the result for the rate of change of the horizontal emittance

$$\frac{d\varepsilon_x}{dt} = \frac{2}{j_x \tau_x} C_q \gamma^2 \frac{I_5}{I_2} - \frac{2}{\tau_x} \varepsilon_x, \tag{3.42}$$

where the first term on the right represents quantum excitation, and the second term represents radiation damping. I_5 is the *fifth synchrotron radiation integral*, given by

$$I_5 = \int_0^{C_0} \frac{\mathscr{H}_x}{|\rho^3|} ds, \tag{3.43}$$

where the function $\mathscr{H}_x$ (sometimes called the 'curly-H' function) is

$$\mathscr{H}_x = \gamma_x \eta_x^2 + 2\alpha_x \eta_x \eta_x' + \beta_x \eta_x'^2. \tag{3.44}$$

The behaviour of the horizontal emittance is similar to that of the energy spread—quantum excitation causes growth in the emittance at a constant rate, independent of the size of the emittance, while the radiation damping leads to exponential decay of the emittance. The quantum excitation is matched by the radiation damping when the horizontal emittance has the value

$$\varepsilon_x = C_q \gamma^2 \frac{I_5}{j_x I_2}. \tag{3.45}$$

The value of the horizontal emittance given by the above expression (3.45) is known as the *natural emittance* of the storage ring, and is usually denoted ε_0. Storage rings for third-generation synchrotron light sources typically operate with natural emittances of order of a few nanometres (see, for example, the facilities mentioned in section 1.3). However, some recent facilities are designed to operate with natural emittance less than 1 nm [13].

A similar analysis to that of radiation effects on the horizontal emittance can be applied to the vertical emittance; however, in a planar storage ring with bending only in the horizontal plane, and in the absence of betatron coupling, there will be no vertical dispersion. Hence, the corresponding quantity to I_5 in the vertical direction will be zero, which suggests that the equilibrium vertical emittance will also be zero. However, in deriving the formula for the natural emittance (3.45), we assumed that synchrotron radiation was emitted purely along the direction of motion of the radiating particle. In fact, as we discussed in section 3.1.3, the radiation is distributed around the direction of motion, with the distribution characterised by an 'opening angle' of $1/\gamma$, where γ is the relativistic factor of the particle. This means that even for a particle moving in the horizontal plane, there is some radiation emitted at (small) angles above and below this plane, which excites vertical betatron motion. Taking this into account, the equilibrium vertical emittance in a ring with no vertical dispersion or betatron coupling is given by [14]:

$$\varepsilon_y = \frac{13}{55} \frac{C_q}{I_2} \int_0^{C_0} \frac{\beta_y}{|\rho^3|} ds. \tag{3.46}$$

In a typical electron storage ring in a third-generation synchrotron light source, this expression gives a value of order 0.1 pm for the vertical emittance. However, even a small amount of vertical dispersion or betatron coupling, from random alignment errors in the magnets, can lead to vertical emittances of order of tens of picometres. Furthermore, collective effects can limit machine performance if the vertical emittance is too small. For example, Touschek scattering, which we shall discuss in section 5.1, causes loss of particles from the beam at a rate that increases as the vertical emittance is reduced. Electron storage rings are typically operated with a vertical emittance that is of order 10 pm or more. In this regime, the vertical emittance is dominated by vertical dispersion and betatron coupling, and the fundamental lower limit from the opening angle of the synchrotron radiation makes only a small contribution.

3.3 Natural emittance and lattice design

For many applications it is important to achieve a small value (of order 1 nm) for the natural emittance in an electron storage ring: this helps to produce synchrotron radiation beams with high brightness in a synchrotron light source, for example. The expression (3.45) derived in the previous section shows that the natural emittance depends on the beam energy (through the factor γ^2) and the ratio $I_5/j_x I_2$. Usually, the beam energy at which a storage ring is operated is determined by considerations such as the wavelength range that the synchrotron radiation should cover: the expression for the critical frequency (3.5) shows that, with the curvature of the beam trajectory limited by magnet technology, a high beam energy is needed to generate high-frequency (short wavelength) synchrotron radiation. Third-generation light sources typically operate with beam energy between 2 GeV and 3 GeV. Achieving low natural emittance, therefore, depends on reducing the ratio $I_5/j_x I_2$—this quantity is determined entirely by the radius of curvature and quadrupole field component in the dipole magnets and insertion devices, and by the optical functions (i.e. the Courant–Snyder parameters and the dispersion). Assuming given field strengths in the dipole magnets and insertion devices, the challenge of achieving a low natural emittance in an electron storage ring must be addressed by appropriate design of the beam optics. In this section, we shall consider a number of different 'styles' of lattice for electron storage rings, and estimate the natural emittance that can be achieved in each case. Further discussion can be found in [15].

3.3.1 FODO lattice

The simplest structure for the magnetic lattice in a storage ring is based on the FODO cell, consisting of alternating focusing (F) and defocusing (D) quadrupoles, with drift spaces (O) or dipole magnets between the quadrupoles. With given magnet strengths and positions, the Courant–Snyder parameters and dispersion can be calculated using the techniques described in chapter 2. An example of the lattice functions in a FODO cell is shown in figure 3.5.

Consider a FODO cell in which the quadrupole magnets have focal length $\pm f$ (positive for the horizontally focusing quadrupole, and negative for the horizontally defocusing quadrupole), and the space between the quadrupoles is occupied by dipole magnets of length L, field strength giving a radius of curvature ρ for the particle trajectory, and no quadrupole field component. In this case, with some approximations (in particular, assuming that $\rho \gg 2f \gg L/2$), it is found that $I_5/j_x I2 \approx 8f^3/\rho^3$, which gives for the natural emittance

$$\varepsilon_{0,\,\text{FODO}} \approx C_q \gamma^2 \left(\frac{2f}{L}\right)^3 \theta^3, \tag{3.47}$$

where $\theta = L/\rho$ is the bending angle of each dipole magnet. To achieve a small natural emittance, we need a small bending angle θ in the dipole magnets, and strong quadrupole magnets to give a small value for f/L.

The number of dipole magnets in the storage ring is $2\pi/\theta$; thus, the smaller the bending angle θ, the more dipole magnets (and the more FODO cells) needed to construct the storage ring, and the more expensive the storage ring will be. The minimum value of θ, therefore, can be limited by costs. There is also a lower limit on the ratio f/L, though from physical rather than financial considerations. The phase advance μ_x in a FODO cell is given by

$$\cos(\mu_x) = 1 - \frac{L^2}{2f^2}. \tag{3.48}$$

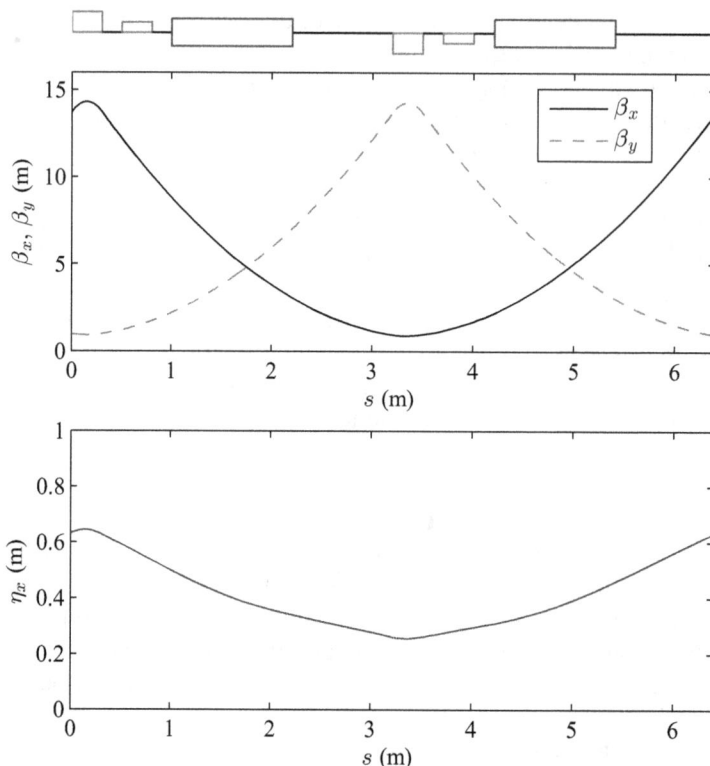

Figure 3.5. Beta functions (top plot) and dispersion (bottom plot) in a FODO cell. In the top plot, the horizontal and vertical beta functions (β_x and β_y) are shown by the solid black and dashed red lines, respectively. The diagram above the plot of the beta functions shows the sequence of magnets in the cell. The long blue rectangles extending above and below the horizontal line represent dipole magnets. Rectangles extending only above the horizontal line represent horizontally focusing quadrupole magnets (taller, red rectangles) and sextupole magnets (shorter, magenta rectangles). Rectangles extending below the horizontal line represent horizontally defocusing quadrupole and sextupole magnets. Note that the horizontal beta function and the dispersion peak in the horizontally focusing quadrupole magnet; the vertical beta function peaks in the horizontally defocusing quadrupole magnet.

Therefore, for stable motion of particles in the beam, the quadrupole magnet strengths and spacing must satisfy

$$\frac{f}{L} \geqslant \frac{1}{2}. \tag{3.49}$$

The minimum value of f/L is $\frac{1}{2}$ (achieved in the limiting case $\mu_x = 180°$), and we therefore expect the minimum natural emittance in a FODO storage ring to be $C_q \gamma^2 \theta^3$. However, for short focal lengths (strong quadrupole magnets) some of the approximations that we have made start to break down, and a more accurate value for the minimum emittance in a FODO storage ring is

$$\varepsilon_{0, \text{FODO, min}} \approx 1.2 C_q \gamma^2 \theta^3, \tag{3.50}$$

which is achieved for a phase advance (in each cell) of approximately 137°.

Usually, accelerators based on FODO cells are designed so that the phase advance per cell is around 90°. This is because the beam optics are less sensitive to focusing (and other) errors at this phase advance than they are at higher phase advances. Assuming a phase advance of 90°, a storage ring consisting of 16 FODO cells (32 dipole magnets) and with a beam energy of 2 GeV would have a natural emittance of approximately 125 nm. This is larger by two orders of magnitude than the emittance needed to achieve the radiation beam brightness required from a modern synchrotron light source. For this reason, light source lattices are now rarely based on FODO cells.

3.3.2 Double-bend achromats

To achieve a smaller emittance than can be obtained in a FODO storage ring, we need to consider a more sophisticated cell structure. One option is the *double-bend achromat* (DBA) [16]. In its simplest form, the central region of a DBA cell consists of a pair of dipole magnets with a quadrupole magnet placed midway between them. The first dipole magnet generates dispersion. The strength of the quadrupole is chosen to reverse the gradient of the dispersion; that is, if the gradient of the dispersion is $\eta_x' = d\eta_x/ds$ at the entrance of the quadrupole, then at the exit of the quadrupole the gradient of the dispersion is $-\eta_x'$. In that case, by the symmetry of the layout, the second dipole magnet exactly cancels the dispersion generated by the first dipole magnet. Hence, in a DBA cell, dispersion is present only in the region between the dipole magnets. The cell is completed by additional quadrupole magnets located outside of the pair of dipole magnets (i.e. in the region with zero dispersion), which can be used to adjust the Courant–Snyder parameters and phase advance to appropriate values. An example of the lattice functions in a DBA cell is shown in figure 3.6.

The benefit of a DBA cell, from point of view of the emittance, is that the dispersion can be kept to a much smaller value in the dipole magnets than is the case in a FODO cell. As a result, the value of $I_5/j_x I_2$ can be smaller than in a FODO cell, for the same dipole magnet parameters. Given the constraints on the dispersion in a DBA cell, it is possible to find values for the Courant–Snyder parameters in the dipoles that minimise the natural emittance. With the appropriate optimisation, the minimum emittance in a DBA storage ring is given by

$$\varepsilon_{0,\,\mathrm{DBA,\,min}} \approx \frac{C_q}{4\sqrt{15}}\gamma^2\theta^3. \tag{3.51}$$

This is smaller than the emittance in a FODO storage ring by a factor of roughly 20, and allows (for a similar size lattice and beam energy) for achieving natural emittances of a few nanometres. A synchrotron light source based on a DBA lattice can reach much higher brightness than a light source based on a FODO lattice.

A further benefit of the DBA lattice style is that it naturally provides zero-dispersion sections at regular intervals around the ring; these sections can be made (in principle) of any desired length, and are ideal locations for insertion devices (see section 3.1.3). If an insertion device is placed at a location with non-zero dispersion,

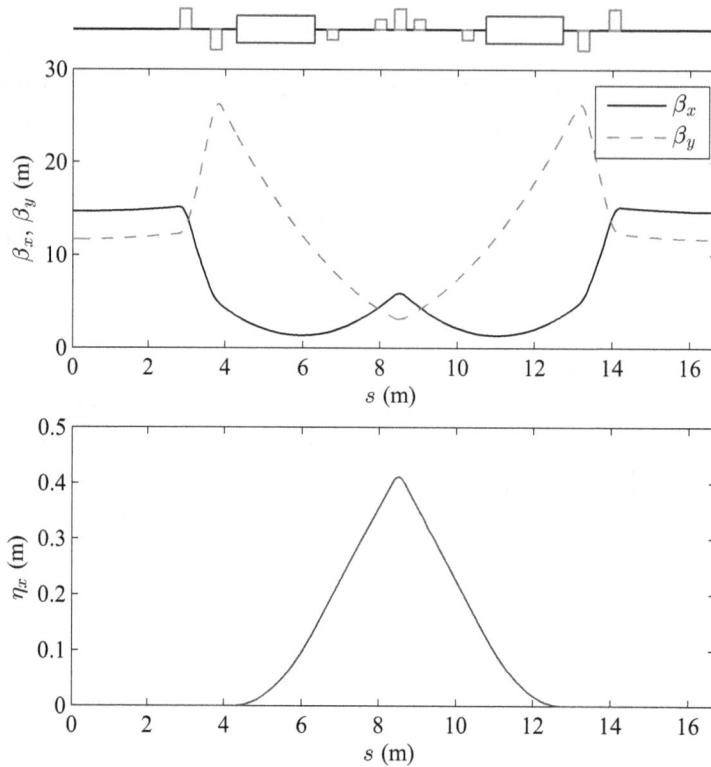

Figure 3.6. Beta functions (top plot) and dispersion (bottom plot) in a double-bend achromat (DBA) cell. In the top plot, the horizontal and vertical beta functions (β_x and β_y) are shown by the solid black and dashed red lines, respectively. The diagram above the plot of the beta functions shows the sequence of magnets in the cell. The long blue rectangles extending above and below the horizontal line represent dipole magnets. Rectangles extending only above the horizontal line represent horizontally focusing quadrupole magnets (taller, red rectangles) and sextupole magnets (shorter, magenta rectangles). Rectangles extending below the horizontal line represent horizontally defocusing quadrupole and sextupole magnets. Note that the quadrupole magnet placed midway between the dipole magnets reverses the gradient of the dispersion, so that the second dipole magnet exactly cancels the dispersion generated by the first.

then quantum excitation in the insertion device leads to an increase in the natural emittance. If the dispersion is reasonably small, this may not be a significant effect, but the large dispersion in a FODO cell can be a drawback from this point of view. A DBA cell, on the other hand, is ideally suited to a low-emittance synchrotron light source needing a large number of insertion devices to serve many users. It is not surprising that DBA cells have been a popular choice for many third-generation synchrotron light sources.

3.3.3 Theoretical minimum emittance lattice and multi-bend achromats

In principle, it is possible to achieve a lower natural emittance than that in a DBA lattice by allowing non-zero dispersion throughout the cell: this removes the constraint that the dispersion (and its gradient) be zero at the entrance of the first dipole magnet in the cell, and at the exit of the second dipole. It is then possible to optimise the Courant–Snyder parameters and the dispersion to minimise the value of $I_5/j_x I_2$ in the cell. The resulting structure is known as a *theoretical minimum emittance* (TME) cell [17]. The natural emittance in a TME storage ring is given by

$$\varepsilon_{0,\,\text{TME}} \approx \frac{C_q}{12\sqrt{15}}\gamma^2\theta^3, \tag{3.52}$$

which is smaller by a factor of 3 than the minimum emittance in a DBA storage ring. However, the TME cell has a number of drawbacks. In particular, dispersion is non-zero throughout a TME lattice, so we lose the advantage of the DBA cell in providing locations with zero (or relatively small) dispersion for insertion devices. Also, nonlinear effects in the beam dynamics (which we shall discuss in chapter 4) can be very strong in a TME lattice, with some adverse consequences for machine performance. For these reasons, the relatively modest benefit of the reduction of the emittance by a factor of three in a TME lattice compared to a DBA lattice is usually outweighed by the drawbacks.

Although the drawbacks associated with the TME cell mean that it has not been a popular choice for synchrotron light sources, it is possible to improve on the performance offered by a DBA lattice by adopting some of the features of a TME cell. One way of doing this is by changing the strength of the quadrupole in the centre of the DBA cell, to allow some dispersion in the (nominally) zero-dispersion regions. Although dispersion now occurs throughout both dipoles, the integral of the curly-H function $\mathcal{H}_x$ (3.44) can actually be reduced in this way, leading to a reduction in the natural emittance; it is now common practice in light sources based on DBA lattices to 'detune' the achromat in this way, to optimise the machine performance. However, depending on the insertion devices installed in the storage ring, if the dispersion is large then quantum excitation in the insertion devices can lead to an overall increase in the natural emittance.

Another way to improve on the DBA cell is to include more dipoles within each cell; the cell is then known as a *multi-bend achromat* (MBA). The advantage of the MBA structure is that it allows the dispersion to be adjusted to zero at either end of the cell, while the dispersion and Courant–Snyder parameters are tuned to be close to the TME conditions in the dipole magnets in the middle of the cell. It is also possible to vary the lengths and bending angles of the dipoles, to optimise the emittance further. It can be shown (see, for example [15]) that in an MBA cell with m dipole magnets in each cell, and with the dipole magnets in the middle of the cell longer than the outer pair of dipole magnets by a factor $\sqrt[3]{3}$, the minimum natural emittance is

$$\varepsilon_{0,\,\text{MBA, min}} \approx \frac{C_q}{12\sqrt{15}}\left(\frac{m+1}{m-1}\right)\gamma^2\theta^3. \tag{3.53}$$

Note that for $m = 2$, this reduces to the expression for the minimum natural emittance in a DBA cell (3.51), and in the limit $m \to \infty$, this reduces to the expression for the natural emittance in a TME cell (3.52). A number of light sources have been constructed with storage rings based on triple-bend achromat (TBA) cells, and some recent facilities use storage rings with significantly larger numbers of dipoles in each cell, to achieve emittances below 1 nm. Examples of light sources based on DBA, TBA, and seven-bend achromat cells are given in table 1.1.

It is worth observing that the above expressions for the minimum natural emittance in different lattice styles (3.50), (3.51), (3.52), and (3.53), all have the general form

$$\varepsilon_0 = FC_q\gamma^2\theta^3, \qquad (3.54)$$

where C_q is the quantum radiation constant, γ is the relativistic factor for the beam, θ is the dipole bending angle, and F is a numerical constant with a value depending on the lattice style (ranging from 1.2 for a FODO lattice, to $1/12\sqrt{15}$ for a TME lattice). The natural emittance scales with the square of the beam energy and with the cube of the dipole bending angle, i.e. as γ^2/N^3, where N is the total number of dipole magnets in the lattice). To achieve a small emittance, therefore, a storage ring should have a low beam energy and a large number of bending magnets; but as we have already observed, there are many other constraints (to do with the radiation properties, beam dynamics, and cost of the facility) that can limit the beam energy and the size of the ring. The design of a lattice for a given facility will usually require some compromise to achieve a reasonable balance between competing effects.

References

[1] Winick H and Doniach S 1980 *Synchrotron Radiation Research* (New York: Plenum)
[2] Willmott P 2011 *An Introduction to Synchrotron Radiation: Techniques and Applications* (Chichester, UK: Wiley)
[3] Mobillo S, Boscherini F and Meneghini C 2015 *Synchrotron Radiation: Basics, Methods and Applications* (Heidelberg, Germany: Springer)
[4] Margaritondo G 2002 *Elements of Synchrotron Light for Biology, Chemistry and Medical Research* (Oxford, UK: Oxford University Press)
[5] Schwinger J 1949 On the classical radiation of accelerated electrons *Phys. Rev.* **75** 1912–25
[6] Sands M 1970 The physics of electron storage rings–an introduction *Technical Report SLAC-121* (Stanford, CA, USA: Stanford Linear Accelerator Center)
[7] Hofmann A 2004 *The Physics of Synchrotron Radiation* (Cambridge, UK: Cambridge University Press)
[8] Rubensson J-E 2016 *Synchrotron Radiation: An Everyday Application of Special Relativity* (San Rafael, California, USA: Morgan & Claypool Publishers) (Bristol, UK: IOP Concise Physics, IOP Publishing)
[9] Clarke J A 2004 *The Science and Technology of Undulators and Wigglers* (Oxford, UK: Oxford University Press)
[10] Jackson J D 1998 *Classical Electrodynamics* III (New York: Wiley)
[11] Shepherd B J A, Scott D J, Hannon F E, Wyles N G and Clarke J A 2005 Commissioning of an APPLE-II undulator at Daresbury Laboratory for the SRS *Proc. 2005 Particle Accelerator Conference (Knoxville, TN, USA)* pp 4051–3
[12] Robinson K W 1958 Radiation effects in circular electron accelerators *Phys. Rev.* **111** 373–80
[13] Borland M 2014 Ultra-low-emittance light sources *Synchrotron Radiat. News* **27**

[14] Raubenheimer T O 1991 The generation and acceleration of low emittance flat beams for future linear colliders *Technical Report SLAC-R-387* (Stanford, CA, USA: Stanford Linear Accelerator Center)

[15] Wolski A 2014 Low-emittance storage rings *Proc. CERN Accelerator School 2014: Advanced Accelerator Physics (Trondheim, Norway, 18–29 August 2013)* ed Werner Herr, CERN–2014–009 (Geneva, Switzerland: CERN) pp 245–94.

[16] Chasman R, Green G K and Rowe E M 1975 Preliminary design of a dedicated synchrotron radiation facility *IEEE Trans. Nucl. Sci.* **NS-22** 1765–7

[17] Lee S Y and Teng L 1991 Theoretical minimum emittance lattice for an electron storage ring *Proc. 1991 Particle Accelerator Conference (San Francisco, CA, USA)* pp 2679–81

Chapter 4

Nonlinear dynamics

As we saw in chapter 2, many of the important features of beam behaviour in a synchrotron storage ring can be understood in terms of the equations of motion of particles moving through drift spaces, dipole magnets, quadrupole magnets, and RF cavities. However, the linear expressions that we used in chapter 2 to describe this motion are only approximations to the full solutions of the equations of motion, which almost always have nonlinear terms. In the components that we have considered so far, the nonlinear terms have relatively weak effects; nevertheless, there are some important phenomena that arise from these terms, and that have significant consequences for the design and operation of electron storage rings. In this chapter, we shall discuss some of the most commonly encountered phenomena associated with nonlinear dynamics in electron storage rings. Generally, the effects of nonlinearities are detrimental, and lead either to rapid loss of particles from the beam, or difficulties in injecting and storing particles in the ring in the first place. In modern storage rings, aiming for high brightness (in light sources) or high luminosity (in colliders), understanding and mitigating the impacts of nonlinearities in the particle motion is essential in the design, commissioning and operation of the machine.

4.1 Chromaticity

The term *chromaticity* refers to the change in the betatron tune with a change in the energy of a particle moving around the ring. In this section, we shall discuss how chromaticity arises in an accelerator beam line, we shall mention some of the potentially damaging effects that chromaticity may have, and finally we shall describe how chromaticity may be controlled so as to avoid any detrimental impact on a storage ring.

doi:10.1088/978-1-6817-4989-1ch4

4.1.1 Natural chromaticity in a storage ring

In section 2.2, we showed that particles moving around a storage ring generally perform transverse oscillations (betatron oscillations) around the closed orbit. The number of oscillations completed in each turn of the ring is called the betatron tune, and is determined by the locations and strengths of the various components (in particular, dipole magnets and quadrupole magnets) around the ring. Particle trajectories are deflected by the field of a quadrupole magnet in much the same way as rays of light are deflected by a lens—a quadrupole magnet acts as a 'magnetic lens' for charged particles. The strength of the focusing from the quadrupole magnets (and the distance between the magnets) is a major factor in determining the betatron tune in a storage ring.

However, for a quadrupole magnet of given strength, the higher the energy of the particle passing through the magnet the smaller the deflection of its trajectory will be (see figure 4.1). In other words, the focal length of the magnetic lens is longer for higher energy particles than for lower energy particles. As a result, higher energy particles will complete fewer betatron oscillations in one turn of a storage ring than lower energy particles: the betatron tune decreases as the particle energy increases. By analogy with a similar effect in light optics (where the focal length of a given lens varies with the colour of the light passing through it), this change in betatron tune with particle energy is called *chromaticity*. Mathematically, the horizontal chromaticity ξ_x in a storage ring is defined by

$$\xi_x = \frac{d\nu_x}{d\delta}, \qquad (4.1)$$

where ν_x is the horizontal tune and δ is the energy deviation (2.18). A similar expression is used to define the vertical chromaticity ξ_y. In principle, the chromaticity in a given storage ring can be calculated from the transfer matrices, since scaling the energy of a particle by a factor $1 + \delta$ has the same effect on the dynamics

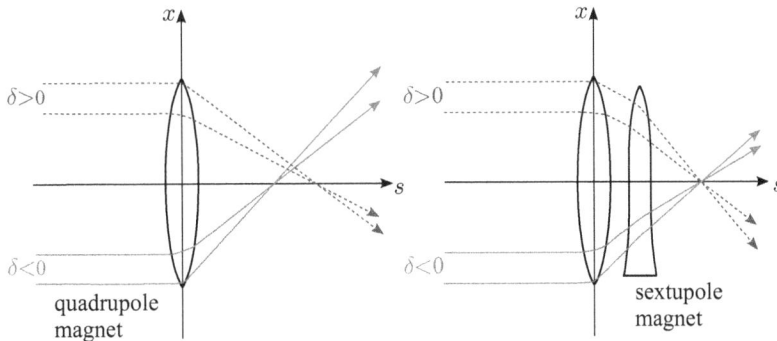

Figure 4.1. Left: the focal length of a quadrupole magnet is shorter for lower energy particles (energy deviation $\delta < 0$) than for higher energy particles ($\delta > 0$). Right: if a sextupole magnet is located where the dispersion is non-zero, then the sextupole provides additional focusing (or defocusing) depending on the energy of the particle. If the strength of the sextupole magnet is chosen appropriately, then the combined system has a focal length that is same for particles with positive, zero, or negative energy deviation.

as scaling the field strengths in each magnet in the storage ring by $(1 + \delta)^{-1}$. It is then possible to calculate the betatron tunes for different values of the energy deviation δ, from which the chromaticities may be obtained. However, it may be shown (see, for example, [1]) that the horizontal chromaticity in a storage ring (including the weak focusing from dipole magnets) is given by

$$\xi_x = -\frac{1}{4\pi} \int_0^{C_0} \beta_x (hk_0 + k_1) \, ds, \tag{4.2}$$

where β_x is the horizontal Courant–Snyder beta function, h is the radius of curvature of the reference trajectory in a dipole with strength $k_0 = (q/P_0)B_y$, and k_1 is the quadrupole focusing strength (2.11), given by $(q / P_0) \, \partial B_y/\partial x$ for a particle with charge q and reference momentum P_0. There is a similar expression for the vertical chromaticity; however, because a quadrupole that focuses a beam horizontally defocuses the beam vertically, and vice versa, the vertical chromaticity is given by

$$\xi_y = \frac{1}{4\pi} \int_0^{C_0} \beta_y k_1 \, ds. \tag{4.3}$$

Since the horizontal beta function β_x is generally larger in horizontally focusing quadrupole magnets than in horizontally defocusing quadrupole magnets, we see from the above formula (4.2) that the horizontal chromaticity resulting from the quadrupole focusing is (as we should expect) negative, i.e. the horizontal tune decreases with increasing energy of the particles. Similarly, if we take into account that the vertical beta function is generally larger in vertically focusing quadrupole magnets (i.e. quadrupole magnets with negative k_1), we see from the above formula (4.3) that the vertical chromaticity resulting from quadrupole focusing is also negative. Calculating the chromaticity using the above formulae is usually easier than working out the betatron tunes for particles of different energies by multiplying the transfer matrices for the components in a beam line.

4.1.2 Correction of chromaticity using sextupole magnets

Chromaticity is an issue in storage rings for two reasons. First, as we already saw in section 2.2.3, the betatron motion of particles moving around a storage ring can be unstable for certain values of the betatron tunes. In particular, for a tune value that is an integer or a half-integer, particles can quickly be lost from the storage ring because of a resonance between the betatron oscillations and the fields from the dipole and quadrupole magnets. We shall see in the next section that there are many other resonances that can have detrimental effects in a storage ring, and the need to avoid the instabilities associated with resonances can place tight constraints on the values that are acceptable for the betatron tunes. However, if the storage ring has a large chromaticity, then the betatron tunes for individual particles can vary significantly as the particles perform synchrotron oscillations. This makes it difficult to find values for the 'nominal' tunes (i.e. the tunes for a particle with zero energy deviation) such that all particles in the beam avoid resonances over the period of a synchrotron oscillation.

The second reason why chromaticity can be an issue in storage rings is that the impact of some collective instabilities can depend on the chromaticity. Collective effects result from the interactions between particles within a beam, and are discussed further in chapter 5. There are many different phenomena associated with collective interactions, and one type of instability in particular, the *head–tail instability*, is sensitive to the chromaticity. In the head–tail instability, electromagnetic fields generated by the head of the bunch drive coherent betatron oscillations in the tail of the bunch. As particles perform synchrotron oscillations, particles in the head and the tail swap longitudinal positions, and the particles now at the head drive oscillations in the tail more strongly, because of their own betatron motion. As a result, the amplitude of the coherent betatron oscillations can grow exponentially. The theory (which we do not discuss in detail here, but which can be found, for example, in [1]) shows that the instability has a growth rate proportional to the magnitude of the chromaticity, and it is only by correcting the chromaticity to a value close to zero that natural damping mechanisms, such as decoherence and synchrotron radiation, can suppress the instability.

In order to operate a storage ring, therefore, it is necessary to correct the chromaticity generated by the quadrupole magnets. This cannot be done by appropriate design of the 'linear' lattice (composed of dipole magnets and quadrupole magnets), since the natural chromaticity in any beam line constructed from dipole magnets and quadrupole magnets will have negative, and possibly quite large, chromaticity. However, it is possible to control the chromaticity using sextupole magnets, in which the vertical field varies with the square of the horizontal distance from the magnetic axis:

$$B_y = \frac{P_0}{q} \frac{k_2}{2} (x^2 - y^2), \tag{4.4}$$

where P_0 is the reference momentum, q is the electric charge on a single particle in the beam, and k_2 is the scaled focusing strength of the sextupole magnet. With the vertical field component given by the above expression, Maxwell's equations require there to be a horizontal field component:

$$B_x = \frac{P_0}{q} k_2 xy. \tag{4.5}$$

In a sextupole magnet, the 'local' field gradient is

$$\frac{\partial B_y}{\partial x} = \frac{P_0}{q} k_2 x. \tag{4.6}$$

The local field gradient determines the focusing strength of the magnet. This can be understood by considering a set of particles travelling parallel to the magnetic axis, with some narrow range of co-ordinates around some value $x = x_0$—the particles experience the same focusing strength as they would in a quadrupole magnet with scaled focusing strength $k_1 = k_2 x_0$. The chromaticity in a lattice can be corrected by adding some extra focusing for particles with a positive energy deviation, and some extra defocusing for particles with a negative energy deviation. In a storage ring, we can take advantage of the fact that the trajectories of particles naturally depend on their energy because of dispersion, so that a particle with energy deviation δ following a closed orbit has x co-ordinate $x_0 = \eta_x \delta$, where η_x is the dispersion. Thus, by placing a sextupole magnet at a location with non-zero dispersion, we provide additional focusing $k_1 = k_2 \eta_x \delta$, which has the required dependence on the energy deviation. This provides a 'correction' for the chromaticity of the quadrupole magnets, as illustrated in figure 4.1.

Sextupole magnets in a lattice can be used to correct the vertical chromaticity as well as the horizontal chromaticity. Including the effects of sextupole magnets, the above expressions (4.2) and (4.3) for the horizontal and vertical chromaticity in a storage ring become

$$\xi_x = -\frac{1}{4\pi} \int_0^{C_0} \beta_x(hk_0 + k_1 - \eta_x k_2) \, ds, \qquad (4.7)$$

and

$$\xi_y = \frac{1}{4\pi} \int_0^{C_0} \beta_y(k_1 - \eta_x k_2) \, ds. \qquad (4.8)$$

To correct the chromaticity for both the horizontal and the vertical motion, we need at least two sets (or *families*) of sextupole magnets: one set with $k_2 > 0$ located where $\beta_x > \beta_y$ (to correct the horizontal chromaticity) and a second set with $k_2 < 0$ located where $\beta_y > \beta_x$ (to correct the vertical chromaticity). Given technical limits on the strengths of the sextupole magnets, the most effective correction of the chromaticity is achieved if the magnets are placed at locations where the dispersion is large, and there is a large difference between the horizontal and vertical beta functions. Hence, the design of the linear optics in a storage ring must take into account the need to correct the chromaticity.

4.1.3 Coupling and nonlinear effects from sextupole magnets

Although sextupole magnets are essential in any storage ring for controlling chromaticity, their use does have some drawbacks. First, the horizontal field component has a dependence on the horizontal distance from the axis of the magnet. This means that a particle passing through the sextupole magnet receives a vertical deflection that depends on its horizontal co-ordinate: hence, sextupole magnets introduce *coupling* (see section 2.6) between the horizontal and vertical motion. Betatron coupling can be an issue for operation of an electron storage ring if achieving low vertical emittance is one of the goals, which is often the case in light sources and in colliders. Fortunately, since the horizontal field component in a sextupole magnet depends on the product of the co-ordinates xy, the strength of the coupling resulting from the sextupole field depends on the vertical offset of the beam from the axis of the magnet. By careful alignment of sextupole magnets during installation, and then by steering the beam during operation so that the closed orbit is as close as possible to the axis of each sextupole magnet, it is possible to reduce the betatron coupling arising from sextupole fields to acceptably low levels.

A more serious drawback associated with the use of sextupole magnets in a storage ring is a consequence of the fact the magnetic fields in a sextupole magnet are nonlinear; that is, the field strength is proportional to products of the horizontal and vertical co-ordinates x and y, rather than simply being proportional to x or y directly (as in a quadrupole magnet). This means, for example, that when a particle passes through a sextupole magnet, it receives a horizontal deflection that depends on the square of its horizontal co-ordinate and the square of its vertical co-ordinate:

$$\Delta p_x \approx \frac{k_2}{2} L (x^2 - y^2), \tag{4.9}$$

where Δp_x is the change in the horizontal momentum of the particle (scaled by the reference momentum), and L is the length of the sextupole magnet. The above formula assumes that the length of the magnet is much less than the betatron wavelength, so that x and y are approximately constant as the particle moves through the magnet.

The fact that the deflections of particle trajectories by sextupole magnets are nonlinear means that the impact of sextupole magnets on the beam optics cannot be analysed using transfer matrices, as we did for 'linear' elements such as dipole magnets and quadrupole magnets. In fact, the beam dynamics resulting from the effects of sextupole magnets (and higher-order multipole fields, such as octupoles, decapoles, etc) are significantly more difficult to analyse and understand than is the case for linear elements. Here, we shall restrict the discussion to two important aspects of the nonlinear dynamics in a storage ring: resonances and dynamic aperture. Both of these aspects are significantly affected by sextupole magnets in a storage ring, and are of major concern at the design stage of an accelerator and during its commissioning and operation. Resonances will be discussed in the next section; dynamic aperture will be discussed in section 4.3 and energy acceptance (closely related to dynamic aperture) will be discussed in the final section in this chapter.

4.2 Resonances

Resonances occur when a particle is repeatedly deflected by a magnetic field each time it passes a particular location in a storage ring, and the deflections add up so as to lead to a steady increase in the amplitude of the betatron oscillations made by the particle. Whether a magnetic field leads to an increase in betatron amplitude depends on the shape of the field (i.e. whether it looks like the field of a dipole magnet, a quadrupole magnet, or a higher-order multipole magnet) and on the betatron tune of the storage ring.

As an example, consider a particle on the closed orbit (i.e. with zero betatron amplitude) in a storage ring. Suppose that there occurs a fault in the power supply for a dipole magnet, so that the strength of the magnet drops slightly. The resulting field can be thought of as a small dipole field error superposed on the nominal field of the dipole magnet. Each time the particle passes through this magnet, it receives a deflection (from the field error) away from the centre of the storage ring. If the horizontal tune is an integer, then on each turn of the ring it becomes deflected further and further from the closed orbit—there is a resonance between the particle motion and the field error that leads to a steady increase in the betatron amplitude of the particle. Eventually, the particle (and all the other particles in the beam) will be kicked out of the storage ring. If the tune is a half-integer, however, then resonance does not occur. Suppose the particle starts on the closed orbit, then receives a deflection from the field error that changes the horizontal momentum by Δp_x. After one complete turn, because the horizontal tune is a half integer, as a result of the betatron oscillations made by the particle, the horizontal momentum will now be $-\Delta p_x$. In the dipole magnet, the particle receives the same deflection as before; but now, because the particle started with a negative horizontal momentum, the horizontal momentum as a result of the deflection will be zero. In other words, over successive turns, the effects of the field error on the particle trajectory will cancel out—there is no resonance, and the particle motion will be stable over the long term.

From point of view of the stability of particle trajectories in the presence of dipole field errors, integer tunes are the worst points at which to operate, while half-integer tunes are optimal. However, consider now the case of a field error on a quadrupole magnet. The deflection of a particle as a result of the error now depends not just on the size of the change in the field gradient in the quadrupole magnet, but also on the horizontal position of the particle as it passes through the magnet. If the horizontal tune is a half-integer, then after each turn both the horizontal co-ordinate and the horizontal momentum change sign. Suppose that on the first turn, the particle has horizontal co-ordinate x and zero horizontal momentum zero in the quadrupole magnet. As a result of the field error, there will be a change in the horizontal momentum, Δp_x. On the next turn the horizontal co-ordinate will be $-x$, and the horizontal momentum will be $-\Delta p_x$. Because of the change in the sign of the co-ordinate, the change in the horizontal momentum on this turn will be $-\Delta p_x$, giving a total horizontal momentum $-2\Delta p_x$. After a further turn, the horizontal momentum will be $3\Delta p_x$, and so on—the horizontal momentum (and betatron amplitude) will continue to increase until the particle is lost from the ring. There is now a resonance between the betatron motion and the quadrupole field error. From the point of view of the stability of particle trajectories in the presence of quadrupole field errors, half-integer tunes are the worst points at which to operate.

Although we have based this discussion of resonances on the effects of field errors, resonances still occur even if a ring is perfectly tuned. In the case of a half-integer resonance, if a particle is on a closed orbit that passes exactly through the centre of every quadrupole, then it can remain on that closed orbit as long as there is no perturbation to its motion. However, any infinitesimal deflection from the closed orbit (arising, for example, from synchrotron radiation) will lead to further deflections from the fields in the quadrupole magnets, that add up from turn to turn to drive the betatron motion of the particle to large amplitude. The particle motion is unstable, even in the absence of any errors in the fields in the magnets.

Higher-order multipole fields can drive resonances for tune values other than integers or half-integers. It might be expected from the preceding argument that sextupole magnets will drive resonances when the tune is a third of an integer, that octupole magnets will drive resonances when the tune is a quarter of an integer, and so on. Although this is true to some extent, the way in which the nonlinear effects of higher-order multipole magnets combine means that the picture is rather more complicated. Pairs of sextupole magnets, for example, can drive not just third-integer resonance, but (in principle) resonances of all orders. An example of the effect of sextupole magnets on the horizontal phase space in a storage ring is shown in figure 4.2; depending on the value of the horizontal tune, the effects of very high-order resonances can be seen. Since any tune value is infinitesimally close to some fraction of an integer, the conclusion may appear to be that there will be some resonance driven for any value of the tune, so particle motion in a storage ring (with higher-order multipoles) can never be stable. However, it is usually the case that the

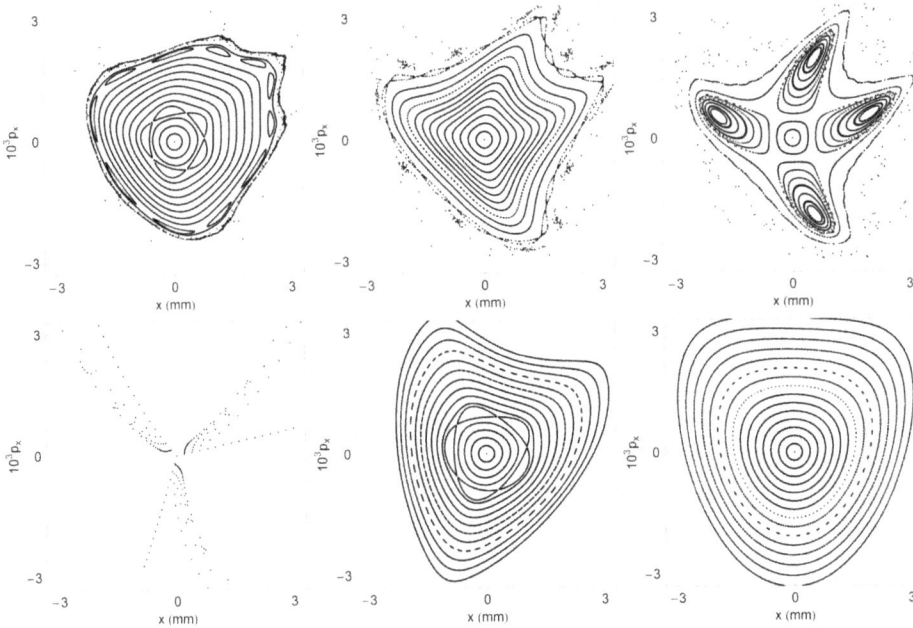

Figure 4.2. Horizontal phase space in a storage ring containing sextupole magnets, as the horizontal tune is varied by changing the quadrupole strengths. Each plot is constructed by tracking a set of particles with different initial conditions over 1000 turns through the storage ring, and plotting the phase space co-ordinates of each particle after each turn. The fractional part of the tune in each plot is (top row, from left to right) 0.202, 0.248, 0.252, and (bottom row, from left to right) 0.330, 0.402, and 0.490. The effect of fifth-order resonances can be seen in the five 'islands' in the plots for tunes 0.202 and 0.402; and the effects of fourth-order and third-order resonances can be seen in the plots for tunes 0.252 and 0.330 (respectively). The third-order resonance leads to unstable particle motion even for very low betatron amplitudes.

higher the order of the resonance, the longer it takes to have any effect, and natural damping mechanisms (in particular, synchrotron radiation) can stabilise the motion. Usually, in electron storage rings, resonances of fifth-order or higher do not significantly affect the stability of particle trajectories—although there can be exceptional cases. In proton storage rings, resonances of much higher than fifth-order can be of concern. The rate at which a resonance drives an increase in betatron oscillation amplitude is usually referred to as the *resonance strength*, and depends on the details of the electromagnetic fields around the storage ring. It is possible for a very high-order resonance, if driven strongly by the fields in the storage ring, to be much more damaging to the stability of particle trajectories than some resonances of lower order.

Of course, resonances occur in the vertical motion as well as in the horizontal motion. In fact, the situation is made still more complicated by betatron coupling, which can lead to resonances occurring if some integer multiples of the horizontal and vertical tunes sum to an integer. The general condition (for betatron resonances) can be written

$$m\nu_x + n\nu_y = p, \qquad (4.10)$$

where ν_x and ν_y are the horizontal and vertical tunes, and m, n, and p are (positive or negative) integers. The order of the resonance is $|m| + |n|$. It is possible to extend the resonance condition to include synchrotron motion—synchro-betatron resonances can impact the stability of particle trajectories, although the synchrotron motion can often be treated (at least, in a first approximation) separately from betatron motion, because of the very different timescales for longitudinal and transverse oscillations in a storage ring.

The resonance condition (4.10) can be illustrated using a *tune grid* (also called a *resonance diagram*, figure 4.3), where the resonance condition is treated as the equation for a set of lines on a graph with axes ν_x and ν_y. Each line corresponds to a different set of values for m, n, and p. A tune grid can be particularly useful for illustrating the impact of effects that change the betatron tunes. The 'nominal' tunes are determined by the dipole and quadrupole magnets in the storage ring lattice; however, the betatron frequencies of a particle can differ from the nominal values because of chromaticity (which describes the change in tunes as a function of the energy of the particle) or nonlinearities (for example, from sextupole magnets) that lead to changes in the magnetic focusing for different betatron amplitudes. Although the nominal tunes, sometimes referred to as the *working point* of the storage ring, can be indicated by a single point on a tune grid, the tunes of the particles in a beam will generally cover some area of the tune grid, known as the *tune spread* or the *tune footprint* of the beam. If the tune spread crosses some low-order resonances on the tune grid, then it may be necessary to try to reduce or control the tune spread, or to try to suppress the effects of the resonances. In general, calculating and controlling the tune spread and mitigating the effects of resonances are complicated tasks; we do not go into them here, but refer the reader to the ample literature on the subject (see, for example [1–5]).

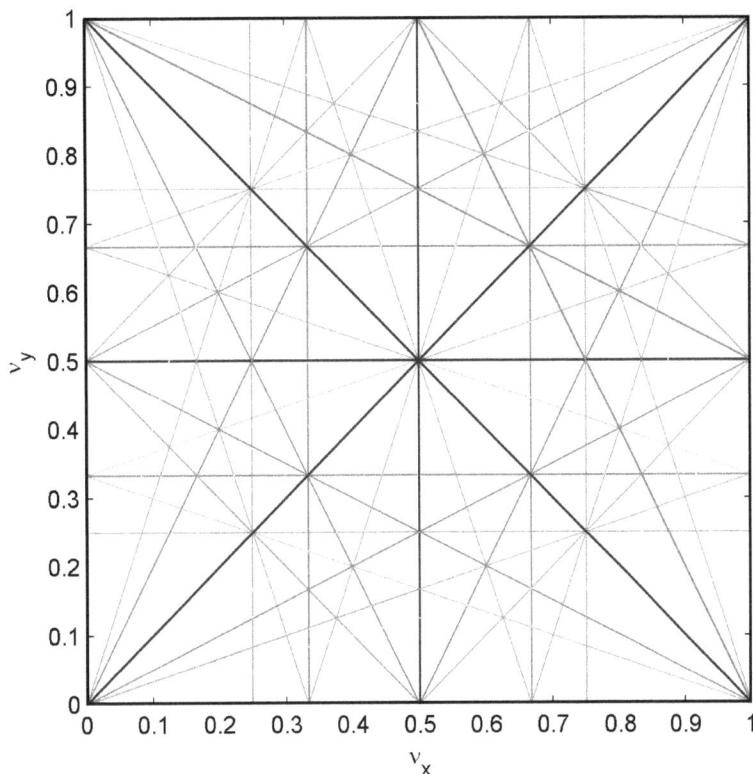

Figure 4.3. Resonance diagram, showing lines in tune space satisfying equation (4.10) for integer values of m, n, and p. Resonance lines up to fifth order (i.e. $|m| + |n| \leqslant 5$) are shown. Blue, red, green, and yellow lines show resonances of second, third, fourth, and fifth order, respectively.

4.3 Dynamic aperture

We have seen in the previous section that resonances between the particle motion and the magnetic fields in an accelerator can lead to particle trajectories being unstable. Although the nominal values of the horizontal and vertical betatron tunes can be chosen to avoid resonances, nonlinear and collective effects can lead to the betatron oscillations of a given particle having tune values differing from the nominal values, depending on factors such as the energy deviation of the particle, and the amplitude of its horizontal and vertical betatron motion. The dependence of the tunes on the energy deviation can be written as a power series; for example, for the horizontal tune

$$\nu_x = \nu_{x0} + \xi_x \delta + \xi_x^{(2)} \delta^2 + \ldots , \tag{4.11}$$

where ν_{x0} is the nominal horizontal tune (for a particle with $\delta = 0$, and zero betatron amplitude), δ is the energy deviation, ξ_x is the linear chromaticity, $\xi_x^{(2)}$ is the second-order chromaticity, and the series continues indefinitely with higher orders in δ. Although the chromaticity can be controlled using sextupole magnets (as discussed in section 4.1.2), usually it is only the linear chromaticity that can be set to zero, and the higher-order chromaticity ($\xi_x^{(2)}$, $\xi_x^{(3)}$, and so on) remains. This means there is generally some residual change in tune with energy deviation, and it is possible that, at large energy deviation, the betatron tunes for a particle can satisfy the resonance condition (4.10) for some particular values of m, n, and p corresponding to a resonance that is strongly driven. In that case, particles performing synchrotron oscillations with sufficiently large amplitude can be lost from the beam.

Even for particles with zero energy deviation, higher-order multipole magnets in the storage ring lattice can lead to changes in tune with betatron amplitude. The situation is analogous to that of a simple pendulum. If the amplitude of the oscillations of the pendulum is fairly small, then the frequency of the oscillations will be (approximately) independent of the amplitude. However, for larger amplitudes, the motion is nonlinear and the frequency of the oscillations will vary depending on the amplitude. In the case of a particle accelerator, the presence of higher-order multipole fields (for example, from sextupole magnets used to correct the chromaticity) will mean that the frequency of the betatron oscillations made by a particle travelling around the ring will depend on the amplitude of those oscillations.

The combination of resonances and changes in the tunes with betatron amplitude means that betatron oscillations at some amplitudes can become unstable. Usually, however, there is some range of amplitudes (in the horizontal and vertical motion) for which the particle motion will be stable. This range is known as the *dynamic aperture* of the storage ring. It is important to achieve a good (large) dynamic aperture in an electron storage ring for a number of reasons. First, particles often have large betatron amplitudes when they are first injected into the ring; if the dynamic aperture is too small it can be difficult, or even impossible, to inject beam into the ring. Second, synchrotron radiation effects lead to particles having some range of betatron (and synchrotron) amplitudes, corresponding to the equilibrium emittances (as discussed in section 3.2.3); if the dynamic aperture is small compared to the equilibrium beam size, then it will not be possible to store beam in the ring for any length of time. Finally, even if the dynamic aperture is sufficient to inject and store the beam, effects such as Touschek scattering (see section 5.1) can lead to some number of particles acquiring very large betatron amplitudes. Although the number of particles with large betatron amplitudes at any given time may be quite small, if the dynamic aperture is such that most of these particles are lost from the beam, then the beam lifetime may be significantly reduced.

The dynamic aperture of a given storage ring depends on the strengths of the resonances in the vicinity of the nominal betatron tunes, and the tune shifts with betatron amplitude. Ultimately, the dynamic aperture is determined by the strengths and locations of the higher-order multipole fields around the storage ring. These fields may be deliberately introduced (for example, in sextupole magnets used to correct the chromaticity), or may occur because of imperfections in the fields of dipole and quadrupole magnets. In electron storage rings in third-generation synchrotron light sources, it is usually the sextupole magnets used to control the chromaticity that limit the dynamic aperture. Optimising the dynamic aperture to ensure good injection efficiency and good beam lifetime can be a difficult task, particularly in low-emittance electron storage rings—low emittance is usually achieved by having low dispersion in the lattice to minimise quantum excitation, but this means that the sextupole magnets need very strong fields in order to correct the chromaticity.

Unfortunately, although there are some general guidelines that can be followed to achieve a good dynamic aperture in a storage ring lattice, there is no single, well-defined procedure that can be applied to reach a given specification. In fact, the complexity of the intrinsically nonlinear motion of particles in a storage ring can make it difficult even to determine the dynamic aperture in a given design. Although some formulae have been developed to provide estimates, usually the best way to find the dynamic aperture is by tracking particles in a computational model of the storage ring. Even then, a trajectory starting from some particular initial conditions may appear to be stable over a large number of turns, only to be lost if the tracking is extended by a few turns more. Furthermore, the dynamic aperture is rarely a region with a well-defined boundary in phase space—two trajectories with very similar initial conditions may eventually exhibit very different behaviour, and it is quite possible for there to be 'islands' of stability that are completely detached from the main dynamic aperture. Finally, the picture is further complicated by phenomena such as Arnold diffusion [6], which describes the possibility for particles to move through phase space, following trajectories that are neither unstable (i.e. outside the dynamic aperture) nor periodic. Ultimately, many of the challenges in calculating the dynamic aperture of a lattice stem from the fact that motion that is strongly nonlinear can be inherently unpredictable. Although many powerful mathematical and computational tools have been developed for the understanding and analysis of nonlinear particle dynamics, the problems involved in fully characterising and then optimising the dynamic aperture in a storage ring have yet to be entirely solved. Further discussion about dynamic aperture, including additional references, can be found in [1, 7].

4.4 Energy acceptance

As mentioned in the previous section, although it is possible to use sextupole magnets in a storage ring to provide some control over the chromaticity, usually it is not possible to eliminate completely the dependence of the tune on the energy of a particle. This means that as particles perform synchrotron oscillations, the frequency of their betatron oscillations will change. Combined with resonances driven by the fields in the various magnets around the ring, the net effect is that the dynamic aperture varies with energy. If the energy deviation is large enough, the dynamic aperture can shrink to zero: the point at which this happens is sometimes known as the *dynamic energy aperture* or the *dynamic energy acceptance* of the storage ring.

Another limitation on the maximum energy deviation for stable oscillations of particles in storage rings comes from the RF voltage. The RF cavities in a storage ring provide the focusing force needed to maintain the stability of synchrotron oscillations; but if the synchrotron oscillation amplitude exceeds a certain value, then a particle will eventually arrive at either a peak or a trough of the RF voltage oscillation. Since the longitudinal focusing strength is related to the gradient of the

RF voltage (i.e. the rate of change of the voltage with respect to time), the longitudinal focusing strength near a peak or a trough of the voltage oscillation is effectively zero, and the RF cavities will not be able to maintain the stability of the synchrotron oscillations. The limit on the energy deviation (for stable synchrotron oscillations) set by the RF cavities is called the *RF acceptance*. The RF acceptance δ_{RF} in a storage ring depends on a number of RF and storage ring parameters, and is given by the formula [1]

$$\delta_{RF} = \frac{2\nu_s}{h|\eta_p|}\sqrt{1 + \left(\phi_s - \frac{\pi}{2}\right)\tan(\phi_s)}, \qquad (4.12)$$

where ν_s is the synchrotron tune, h is the harmonic number, η_p is the phase slip factor, and ϕ_s is the synchronous phase.

Finally, limits on the maximum energy deviation of particles in a storage ring can come from physical apertures. If the dispersion at a particular point in a storage ring is 1 m, then a particle with 1% energy deviation at that point will follow a trajectory that is 1 cm away from the closed orbit. If the beam pipe is not wide enough to allow for such large trajectory deviations, then the particle will hit the wall of the beam pipe and will be lost from the machine.

A particle will not remain in the beam if its energy deviation is larger than the dynamic energy acceptance, the RF acceptance or the physical acceptance (set by the beam pipe aperture). Typically, the dynamic energy acceptance in a storage ring for a third-generation synchrotron light source is a few percent. The higher the RF voltage, the larger the RF acceptance will be; however, there is little benefit in providing an RF acceptance that is much larger than the dynamic energy acceptance. Similarly, since there are costs associated with providing a large beam pipe aperture, there is little advantage in making the physical acceptance larger than either the dynamic energy acceptance or the RF acceptance.

Typically, the equilibrium energy spread (rms energy deviation) in an electron storage ring is of order 0.1%, so an energy acceptance of a few percent would appear to be more than sufficient to be able to store a beam. This is indeed the case; however, even when an energy acceptance of several percent is achieved, a further increase in the acceptance can have operational benefits by improving injection efficiency and beam lifetime. Regarding injection, the beam injected into a storage ring can often include particles with large energy deviation, and making the energy acceptance in the storage ring large enough to accept all the particles in the injected beam will minimise injection losses. With regard to beam lifetime, a large energy acceptance helps to minimise the rate of loss of particles from Touschek scattering (discussed further in section 5.1). In a low-emittance electron storage ring, Touschek scattering between particles in a bunch leads to a transfer of the transverse momentum of a particle to longitudinal momentum, with the result that the energy deviation of a particle after a scattering event can be very large; if the energy deviation is outside the energy acceptance of the ring, the particle will be lost from the beam. This results in a continuous drop in beam current at a rate characterised by the *Touschek lifetime*, which is typically a few hours. Operational limitations imposed by a short Touschek lifetime can be overcome by injecting small amounts of beam current at short, regular intervals (perhaps a few minutes), but 'top-up injection' [8] has a number of implications for the design and operation of a storage ring and improvement of the Touschek lifetime, where possible, is often beneficial.

There are a number of reasons, therefore, why it is desirable in a storage ring to achieve an energy acceptance that is as large as possible, and it is often essential to achieve an acceptance that is larger than the equilibrium rms energy spread by at least an order of magnitude. The dynamic energy acceptance is often the main limitation; however, as is the case with the dynamic aperture, the fact that the dynamic energy acceptance is determined by nonlinear effects makes it difficult to optimise or even characterise this quantity. At the design stage, the energy acceptance can be estimated by tracking particles in a computational model of the storage ring (in a similar process to that used to estimate the dynamic aperture). When a storage ring is operational, the dynamic energy acceptance can be found by measuring the beam lifetime as a function of the RF voltage (see, for example, [9]). Assuming that the beam lifetime is dominated by Touschek scattering, increasing the voltage from a low level (at which the overall energy acceptance is limited by the RF acceptance) will increase the beam lifetime. However, at some point the energy acceptance becomes dominated by the dynamic energy acceptance, and increasing the RF voltage beyond this point will lead to no further improvement in beam lifetime—in fact, the beam lifetime will actually decrease, since increasing the RF voltage reduces the bunch length, which increases the particle density in a bunch and hence increases the rate of Touschek scattering.

References

[1] Wolski A 2014 *Beam Dynamics in High Energy Particle Accelerators* (London, UK: Imperial College Press)

[2] Herr W 2014 Mathematical and numerical methods for non-linear beam dynamics *Proc. CERN Accelerator School 2013: Advanced Accelerator Physics (Trondheim, Norway, 18–29 August 2013)* ed W Herr (Geneva, Switzerland: CERN), 157–98 number CERN-2014-009 (arXiv:1601.07311)

[3] Wiedemann H 2015 *Particle Accelerator Physics* 4th edn (New York: Springer)

[4] Forest É 1998 *Beam Dynamics: A New Attitude and Framework* (Amsterdam, Netherlands: Harwood Academic)

[5] Dragt A J 2017 Lie methods for nonlinear dynamics with applications to accelerator physics http://www.physics.umd.edu/dsat/dsatliemethods.html [Online; accessed 3 January 2018]

[6] Arnold V I 1964 Instability of dynamical systems with several degrees of freedom *Sov. Math.* **5** 581–8

[7] Wu Chao A, Mess K H, Tigner M and Zimmermann F (ed) 2013 *Handbook of Accelerator Physics and Engineering* 2nd edn (Singapore: World Scientific)

[8] Ohkuma H 2008 Top-up operation in light sources *Proc. 2008 European Particle Accelerator Conference (Genoa, Italy)* pp 36–40

[9] Steier C, Robin D, Nadolski L, Decking W, Wu Y and Laskar J 2002 Measuring and optimizing the momentum aperture in a particle accelerator *Phys. Rev. E* **65** 056506

Chapter 5

Collective effects

Particles within a beam in a storage ring can interact with each other in various ways. For example, a bunch of charged particles can generate electromagnetic fields, known as *wake fields*, within a section of the vacuum chamber; these fields can persist for some time after the bunch has passed by, and can exert forces on following bunches. Also, particles within a bunch can 'collide' as they perform betatron and synchrotron oscillations while moving around the ring. Particles within a beam can further interact through the build-up of charged particles in the vacuum chamber, created by ionisation of residual gas or dust particles; these charged particles can exert forces on particles in the beam, affecting their dynamics. Although varied in nature, these effects have the common property that their impact depends on the amount of current in the beam.

Generally, phenomena associated with the interactions between particles in an accelerator are known as *collective effects*, and are manifest in various ways. In an electron storage ring, the impact of a particular kind of interaction may be limited to some change in the equilibrium distribution of particles within a bunch; for example, the emittance or energy spread may be observed to increase with the total charge of the bunch. Some collective effects may have a more significant impact, in preventing a bunch from reaching an equilibrium distribution. In such cases, the collective effect is known as a *beam instability*. Ultimately, collective effects may lead to the loss of particles from the beam, with a significant drop in beam current occurring on timescales ranging from several hours to a fraction of a second, depending on the loss mechanism. Modern storage rings often require high beam currents to achieve their performance specifications. Understanding and (where possible) mitigating collective effects is then an essential part of the design and operation of the storage ring. In this chapter, we shall consider some of the principal collective effects commonly encountered in electron storage rings. These include scattering effects (from collisions between particles) leading to limitations on the beam lifetime, and the effects of wake fields that often result in beam instabilities.

5.1 Touschek scattering and space charge

We begin by considering the scattering processes that occur when two particles within a bunch collide. One effect of this scattering, as we shall see, is the loss of particles from the beam. The limit that this imposes on the beam lifetime is an important aspect of the operational performance of many modern electron storage rings.

As a bunch of particles moves around a storage ring, particles within the bunch perform betatron and synchrotron oscillations. Occasionally, in the course of these oscillations, two particles may come close enough to each other that the force between them is sufficiently large to cause a significant change in their motion. In the case of electrons, such an interaction is described by *Møller scattering* [1]. The theory of Møller scattering gives the differential cross-section that describes the probability for two electrons to be deflected by a certain angle following a collision. Suppose, for example, that two electrons move towards each other along trajectories that are parallel to the x (horizontal, transverse) axis, but with some small separation. As they move past each other, the electrons can be deflected so that there is some vertical and longitudinal component to their motion. If the deflection angle is fairly small, the only consequence may be some transfer of momentum between the different degrees of freedom of the particle motion. This may be observed in a storage ring as a change in the beam emittances, in a process known as *intrabeam scattering* (IBS) [2, 3]. Although IBS takes place continually in bunches consisting of large numbers of particles, the rate of change of the emittances that happens as a result of IBS in an electron storage ring is usually slow compared to the synchrotron radiation damping times. This means that the effects of IBS in electron rings are generally difficult to observe, and only become significant under special conditions of high particle density [4–6]. If the particle density is increased by increasing the number of particles in a bunch, then other collective effects can start to impact the dynamics, masking the effects of IBS; usually the best way to observe IBS in an electron storage ring is to reduce the vertical emittance to as low a value as possible.

A more important effect occurs when two particles in a bunch approach each other closely enough that they are deflected through an angle leading to a significant transfer of momentum from a transverse to the longitudinal direction. Since the betatron tunes are generally much larger than the synchrotron tune, in the rest frame of the bunch the transverse component of the momentum will be larger than the longitudinal component, for most of the particles in the bunch. This means that scattering between particles can lead to a significant increase in the longitudinal momentum of the particles; it is then possible that following a scattering event, the energy deviation of one or both of the particles involved will be outside the energy acceptance of the storage ring (as discussed in section 4.4). In that case, the particles will be lost from the beam, and the process is known as *Touschek scattering* [7]. The observable effect is that the beam current in the storage ring decays over time. The rate of decay depends on the beam energy, the beam size (transverse and longitudinal) and the bunch population. In a parameter regime typical for a third-generation synchrotron light source, the Touschek lifetime characterising the decay rate is usually a number of hours. Other processes (for example, the scattering of electrons from residual gas molecules in the vacuum chamber) also lead to the loss of particles from the beam, but at low emittance and high bunch charge, Touschek scattering normally dominates the beam lifetime.

The rate of loss of particles from a beam by Touschek scattering is proportional to the square of the number of particles in a bunch; strictly speaking, this means that the beam current does not decay exponentially. However, for a short time interval (over which the bunch population is roughly constant) the fall in current can be approximated as an exponential decay, so that

$$I(t) = I(0)e^{-t/\tau_\mathrm{T}}, \tag{5.1}$$

where $I(0)$ is the initial beam current, $I(t)$ is the beam current at time t. The *Touschek lifetime* τ_T is given by

$$\tau_\mathrm{T} = \frac{N_\mathrm{b}cr_\mathrm{e}^2}{8\pi\gamma^2\sigma_x\sigma_y\sigma_z}\left(\frac{D(\xi)}{\delta_\mathrm{max}}\right)^3, \tag{5.2}$$

where N_b is the bunch population, r_e is the classical radius of the electron[1], γ is the relativistic factor for particles in the bunch, σ_x, σ_y, and σ_z are the horizontal, vertical, and longitudinal rms beam sizes, δ_max is the energy acceptance of the ring, and $D(\xi)$ is the function

$$D(\xi) = \xi^{\frac{3}{2}}\int_\xi^\infty \frac{e^{-u}}{u}\left(\frac{u}{\xi} - 1 - \frac{1}{2}\ln\left(\frac{u}{\xi}\right)\right)du. \tag{5.3}$$

The parameter ξ is defined by

$$\xi = \frac{\delta_\mathrm{max}^2\beta_x}{\gamma^2\varepsilon_x}, \tag{5.4}$$

where β_x is the horizontal Courant–Snyder parameter, and ε_x is the horizontal emittance. The Touschek function $D(\xi)$ is plotted in figure 5.1.

Particles in a beam in an accelerator can be affected not just by individual scattering events, but also by interaction with the electromagnetic field generated collectively by all the particles in a bunch. If we consider a bunch of electrons at rest in the laboratory, the electrostatic repulsion between the electrons will cause the bunch to expand rapidly—the bunch will not survive very long. In an accelerator, relativistic effects act in two ways to reduce the rate of expansion of a bunch travelling at close to the speed of light. First, Lorentz contraction means that a bunch that appears relatively short (of order of a few millimetres) when moving in an accelerator will be longer by the relativistic factor γ in the rest frame of the bunch,

---

[1] For electrons, with charge e and mass m_e, the classical radius is $r_\mathrm{e} = e^2/4\pi\epsilon_0 m_\mathrm{e}c^2 \approx 2.817\ 94 \times 10^{-15}$ m.

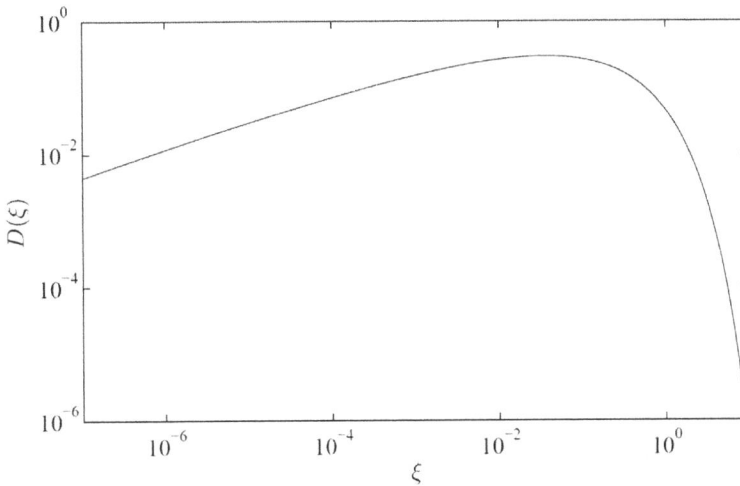

Figure 5.1. The Touschek function $D(\xi)$, defined by equation (5.3). The parameter ξ (5.4) increases with the energy acceptance of the storage ring, and decreases with the beam energy and the emittance. The Touschek lifetime is proportional to the third power of $D(\xi)$.

with the result that the charge density will be reduced by the same factor. Second, time dilation means that the expansion rate of the bunch in the accelerator is reduced by a factor γ when observed from the laboratory rest frame. Overall, the expansion rate is reduced by a factor γ^2 compared to a bunch with the same dimensions observed in its rest frame. For a beam at 2 GeV in an electron storage ring, γ is roughly 4000, so the suppression of the effects of the Coulomb repulsion by a factor γ^2 is significant.

Another way of looking at the (effective) suppression of the Coulomb repulsion is by considering the fields around a bunch of particles moving through an accelerator. The moving charged particles constitute an electric current, which generates a magnetic field around the bunch. A particle within the bunch sees both the electric field and the magnetic field. Although the force from the electric field acts to push the particle out of the bunch, the force from the magnetic field is such as to pull the particle towards the bunch. A detailed analysis shows that overall, the transverse acceleration of a particle resulting from the combined electric and magnetic fields is reduced by a factor γ^2 in comparison to the acceleration that would be expected for a bunch of particles at rest (with no magnetic field). This is consistent with the same result obtained by considering the effects of time dilation and Lorentz contraction.

Although the Coulomb repulsion is suppressed in an accelerator, the fact that the relativistic factor γ is always a finite quantity means that there are always some residual effects from this repulsion, these are generally known as *space charge effects*. In the transverse directions, space charge forces may be viewed as defocusing forces, since the size of the force varies with distance from the centre of the bunch: space charge forces can then affect the tunes and the Courant–Snyder parameters in a storage ring. However, because of the way in which charge is usually distributed within a bunch, the defocusing force is strongly nonlinear, and different particles will experience different defocusing strengths depending on their transverse and longitudinal position within the bunch. Space charge then leads to a spread of different tune values for different particles, sometimes called an *incoherent tune shift*. However, since the relativistic factor is typically quite large in electron storage rings, space charge effects are usually weak, and although they may be significant in certain parameter regimes, we do not consider space charge effects further here. Further discussion can be found in, for example, [8–11].

5.2 Ion trapping

Storage rings must achieve extremely low residual gas pressures within the beam pipe (or vacuum chamber) for two main reasons. First, particles in the beam can scatter from gas molecules in the beam pipe and be lost from the beam as a result— this can limit the beam lifetime. Second, gas molecules can be ionised by the electromagnetic fields around the beam, and the ions produced in this way can interact with the beam, potentially causing it to become unstable. Pressures of order of 1 nTorr are routinely achieved in electron storage rings; for comparison, atmospheric pressure is roughly 760 Torr.

Ion effects [12, 13] can be particularly damaging in a storage ring if the ions build up to high densities. The forces exerted by the ions can deflect the trajectories of bunches in the beam; the beam in turn deflects the ions, and it is possible for a cycle to develop in which the beam starts to perform oscillations of increasing amplitude, leading eventually to the loss of beam current. Beam loss is most likely to occur if ions become trapped in the beam, reaching high densities as ions continue to be generated from the residual gas molecules in the vacuum chamber. Ion trapping can be understood in terms of the focusing effects of the fields around bunches in the beam—the electric and magnetic fields generated by a bunch leads to forces on the ions that increase with transverse distance from the centre of the bunch. In that respect, the forces are similar to those from quadrupole magnets acting on particles in the beam; however, the forces experienced by ions from the fields around the beam are focusing both horizontally and vertically, and are also strongly nonlinear. Despite the nonlinear nature of the forces, it is possible to analyse the dynamics of ions in the beam in much the same way as the dynamics of particles in the beam can be analysed as the beam moves through (for example) a FODO beam line. In both cases, particles receive regular 'kicks' separated by drift lengths.

In the case of particles travelling along a FODO beam line, there are constraints on the quadrupole strengths and separations for the particle motion to be stable. There are corresponding constraints on the focusing forces experienced by ions and the bunch separation for the motion of the ions to be stable in the presence of a beam. The stability condition can be expressed in terms of the mass to charge ratio A/Q of the ions (with A the relative atomic mass, and Q the charge of the ion in units of the elementary charge e). Ions can be trapped (perform stable oscillations) in the beam if [13]

$$\frac{A}{Q} \geqslant \frac{N_b r_p L_{\text{sep}}}{2\sigma_y(\sigma_x + \sigma_y)}, \tag{5.5}$$

where N_b is the number of particles in each bunch in the beam, bunches have regular separation L_{sep}, r_p is the classical radius of the proton, and σ_x and σ_y are the horizontal and vertical rms beam sizes, respectively. It is assumed that $\sigma_y < \sigma_x$, which is usually the case in an electron storage ring.

In many cases, it is inevitable that some species of ion generated from residual gas molecules in the vacuum chamber will be trapped in the beam. Unfortunately, improving the gas pressure cannot help, since gas molecules cannot be entirely removed from the chamber, and the ion density will eventually build up to sufficiently high levels as to cause beam instability. An effective mitigation, however, is to include a gap in the bunch train, of sufficient length to disrupt the focusing effect of the beam and destabilise the motion of the ions. Although this means that some beam current must be sacrificed because a number of the RF buckets will be empty, the benefit in terms of beam stability is usually essential for meeting the performance required of the storage ring.

5.3 Wake fields, wake functions, and impedances

The electromagnetic fields generated by charged particles in the beam in an accelerator are modified by the presence of the vacuum chamber and components such as RF cavities, beam position monitors, flanges (where sections of beam pipe are joined together), and bellows (that allow for some expansion or movement of the beam pipe). As particles pass a given section of beam pipe, they can excite oscillating electromagnetic fields in that section that may persist for some time after the bunch has passed by. These fields, known as *wake fields*, can exert forces that affect the motion of other particles within the accelerator [14, 15]. Wake fields are sometimes categorised as *short range* or *long range*. Short range wake fields are those that are generated and act back on particles within a single bunch. In an electron storage ring, bunch lengths are typically a few millimetres. Long range wake fields persist for much longer, and act on bunches following the bunch originally generating the wake field; the length scale in this case can be anything from around a metre, to some hundreds of metres. An example of the wake fields in a section of accelerator beam pipe generated by an electron bunch is shown in figure 5.2.

Calculating the wake fields as functions of position and time from a bunch with given dimensions and charge, and in a section of beam pipe with given geometry and materials, is a complex computational problem, usually involving numerical solution of Maxwell's equations [16, 17]. The task can be computationally expensive, because of the range of distance scales involved. The bunch generating the wake fields may be a few millimetres long, but much less than a millimetre in width and height. This implies the need to model the fields on a grid (in space and time) with a spacing between grid points of a fraction of a millimetre. However, the grid may need to cover a volume with dimensions of several centimetres transversely and up to a metre longitudinally. Assuming that the wake field in a given situation can be computed, determining its impact on the beam in an accelerator is another formidable challenge. There is again a very wide range of distance and time scales

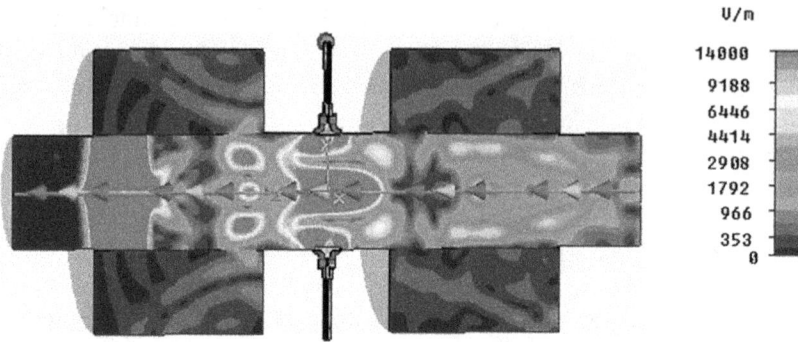

V/m
14000
9188
6446
4414
2908
1792
966
353
0

Figure 5.2. Wake fields in a section of accelerator beam pipe containing a beam position monitor (BPM). A cross-section (in the vertical–longitudinal plane) of the beam pipe is shown, with the beam moving from right to left. The colours show the strength of the electric field generated by a bunch of electrons: the bunch is located in the large red area (high electric field strength) towards the left of the diagram. The pick-up electrodes for the BPM are in the centre of the section of beam pipe, with bellows (to allow some flexibility in the beam pipe position on either side of the BPM) on the left and right of the electrodes. Copper strips (not visible in the diagram) are positioned to prevent the electric fields penetrating into the bellows.

involved, and it is usually not feasible to compute the motion of each individual particle even within a single bunch, which may contain of order 10^{10} particles.

There are also numerous ways in which wake fields may affect beam behaviour. In some cases, there may simply be some increase in the transverse or longitudinal dimensions of a bunch, but with the shape of the bunch (usually, in an electron storage ring, described by a Gaussian distribution function) remaining the same; or, the shape of the bunch may change as it moves around the ring. An example of the impact of wake fields on the longitudinal phase space charge density in a single bunch in a storage ring is shown in figure 5.3. As well as changing the charge distribution with a bunch, wake fields can change the betatron or synchrotron tunes, in which case the effects can depend on the proximity of the nominal tunes to resonances. Entire bunches may start to oscillate, either transversely or longitudinally. Finally, the problem is further complicated by the fact that as the beam responds in some way to the presence of wake fields, the wake fields that it continues to generate will change, because of changes in the size, shape or trajectory of the bunches. For an accurate description of beam behaviour in a storage ring with wake fields, it may be necessary to find a self-consistent solution to the full system including the generation of the wake fields and the dynamics of the beam in the presence of the wake fields.

Given the complexity of the problem, it is not surprising that we look for simplified models that can nevertheless give us some insight into the beam behaviour, and have some value not only in predicting how a beam will be affected by wake fields in a given accelerator design, but also how the design may be optimised in order to achieve the best possible performance. The first simplification is to represent the wake fields by *wake functions* [14], which describe the change in energy or momentum of one particle as it follows another particle through a given

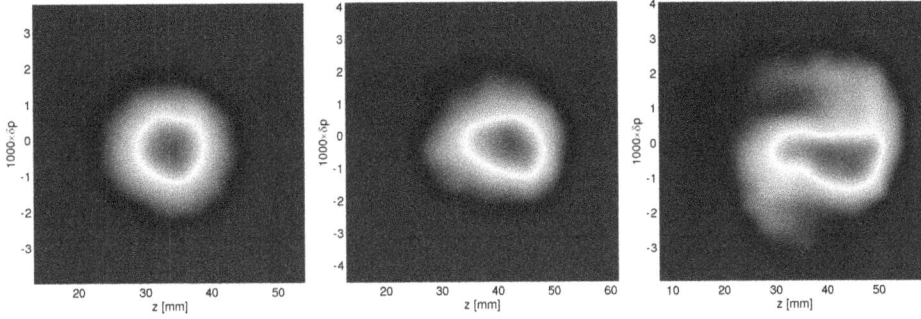

Figure 5.3. Simulation of the effect of wake fields on the charge distribution in longitudinal phase space of a single bunch in a storage ring. At low charge (left-hand image) the bunch distribution is Gaussian, and stable. At higher charge (middle image) some distortion is visible on the distribution. If the bunch charge is increased still further (right-hand image) the distortion becomes more pronounced. In this case, the distribution fails to reach an equilibrium, but continuously changes shape.

section of the accelerator beam line. In particular, we can define a longitudinal wake function $W_{\parallel}(-\Delta z)$ with units of V C^{-1} (volts per coulomb), such that

$$\Delta \delta_B = -\frac{q_A q_B}{E_0} W_{\parallel}(-\Delta z), \qquad (5.6)$$

where the wake field is generated by a particle (or small bunch of particles) with charge q_A and longitudinal co-ordinate z_A, and a trailing particle with charge q_B and longitudinal co-ordinate z_B undergoes a change in energy deviation $\Delta \delta_B$ over the length of the given accelerator section. E_0 is the reference energy and $\Delta z = z_A - z_B$. Note that Δz is positive for $z_A > z_B$, i.e. if particle A is ahead of particle B—causality implies that $W_{\parallel}(-\Delta z) = 0$ if Δz is negative.

Similarly, we can define a transverse wake function $W_{\perp}(-\Delta z)$ with units of V C^{-1} m^{-1}, which gives the transverse deflection of particle B as it follows particle A through a given section of the accelerator:

$$\Delta p_{y, B} = -\frac{q_A q_B}{E_0} y_A W_{\perp}(-\Delta z). \qquad (5.7)$$

Here, $p_{y, B}$ is the vertical momentum of particle B scaled by the reference momentum, and y_A is the vertical co-ordinate of particle A. This formula assumes that the transverse deflection of particle B has a linear dependence on the transverse position of particle A. This may be a good approximation over a small range of transverse offsets from the reference trajectory (which we assume coincides with the centre of the beam pipe), but will break down at some point.

There are two main benefits of working with wake functions rather than wake fields. First, the wake field depends on the properties (charge, size and shape, trajectory) of the bunch generating the wake field as well as on the properties (geometry, materials) of the section of beam pipe under consideration. Using a wake function allows us to separate the properties of the bunch generating the wake fields from the properties of the beam pipe—a wake function describes a section of the accelerator, regardless of the beam passing along the accelerator. The second benefit is that wake functions reduce a wake field that is a complex function of position (in three dimensions) and time, to a relatively simple function that depends only on one variable, namely the longitudinal separation between a point-like particle generating a wake field, and a point-like particle observing the wake field. Both benefits, of course, come at the sacrifice of some accuracy and completeness in the description of the system. In many cases, however, the use of wake functions rather than wake fields makes it possible to gain some insight into the behaviour of a system (a beam in an accelerator) that would otherwise be hopelessly complicated.

In some cases, wake functions can be calculated analytically by finding a solution to Maxwell's equations for the fields around a point charge moving close to the speed of light, with the boundary conditions imposed by the vacuum chamber [14, 18]. For example, in the case of a long, straight vacuum chamber of length L and with uniform circular cross-section of radius r, the longitudinal wake function is given (for $\Delta z > 0$) by

$$W_{\|}(-\Delta z) = \frac{1}{2\pi r}\sqrt{\frac{Z_0 c^2}{4\pi\sigma}}\frac{L}{\sqrt{\Delta z^3}},\tag{5.8}$$

where Z_0 is the impedance of free space, and σ is the conductivity of the vacuum chamber material. The transverse wake function is given (again for $\Delta z > 0$) by

$$W_{\perp}(-\Delta z) = -\frac{2}{\pi r^3}\sqrt{\frac{Z_0 c^2}{4\pi\sigma}}\frac{L}{\sqrt{\Delta z}}.\tag{5.9}$$

Since the wake fields in this case arise essentially from the finite conductivity of the vacuum chamber, these wake functions are known as the *resistive wall* wake functions. By causality, both wake functions vanish if the particle observing the wake fields is ahead of the particle generating the wake fields, i.e. if $\Delta z < 0$. It should also be noted that the above expressions involve approximations that are valid if

$$\Delta z \gg \sqrt[3]{\frac{r^2}{Z_0\sigma}}.\tag{5.10}$$

Although analytical calculations are possible in some cases, wake functions generally need to be computed the same way as wake fields, by solving Maxwell's equations numerically in a given section of beam line; unfortunately, the computation of a wake function can be more difficult than that of a wake field, since the source of the fields should ideally be represented as a point-like particle. This raises certain computational difficulties, which we do not discuss further. For an overview of methods used for computing wake functions, see (for example) [17] and references therein.

Wake functions provide a description of wake fields in the time domain. A wake function is a function of the longitudinal separation of two particles (where the longitudinal separation is specified in terms of the co-ordinate z which, it should be remembered, actually gives the time that a particle arrives at a given point in the beam line). It turns out that in many cases, the most damaging wake fields are associated with resonances in the beam pipe, where fields oscillating at some specific frequency (the resonant frequency) can persist for long periods of time. In such cases, it is often more convenient to work with a description of the wake fields in the frequency domain, in which case we express the wake function as a superposition of sinusoidal functions of different frequencies. In other words, we work with the Fourier transform of the wake function, which is known as the *impedance* [14, 15, 19]. The longitudinal impedance is defined by

$$Z_\|(\omega) = \int_{-\infty}^{\infty} W_\|(z) e^{-i\frac{\omega z}{c}} \frac{dz}{c}, \tag{5.11}$$

and the transverse impedance is defined by

$$Z_\perp(\omega) = i \int_{-\infty}^{\infty} W_\perp(z) e^{-i\frac{\omega z}{c}} \frac{dz}{c}. \tag{5.12}$$

With these definitions, the longitudinal wake function can be expressed in terms of the longitudinal impedance by an inverse Fourier transform:

$$W_\|(z) = \frac{1}{2\pi} \int_{-\infty}^{\infty} Z_\|(\omega) e^{i\frac{\omega z}{c}} d\omega. \tag{5.13}$$

Similarly, for the transverse wake function

$$W_\perp(z) = -\frac{i}{2\pi} \int_{-\infty}^{\infty} Z_\perp(\omega) e^{i\frac{\omega z}{c}} d\omega. \tag{5.14}$$

Note that the longitudinal impedance has units of Ω (ohms), while the transverse impedance has units of Ωm^{-1} (ohms per metre).

Simply knowing the form of an impedance for a section of accelerator beam line can be useful in providing some indication of whether the wake fields are likely to have a strong impact on the beam. This is because the effect of the wake field depends on the convolution of the beam spectrum (i.e. the frequencies present in the beam current observed at a particular location as a function of time) and the impedance. In particular, it can be shown (see, for example [10]) that

$$\tilde{V}(\omega) = \tilde{I}(\omega) Z_\|(\omega), \tag{5.15}$$

where $\tilde{V}(\omega)$ is the Fourier spectrum of the voltage (seen by a particle in the beam) across a section of beam line with impedance $Z_\|(\omega)$, and $\tilde{I}(\omega)$ is the beam current spectrum. If the beam current has a component at a frequency where the impedance is large (perhaps associated with a resonance), then there will be a large voltage induced across that section of the beam line, leading to a large change in energy of the particles in the beam as they pass through. This could have a significant impact on the beam behaviour, potentially leading to an instability. On the other hand, if the impedance is small at the frequencies that dominate the beam current spectrum, the wake fields are unlikely to have any great impact on the beam. The above equation (5.15) is, of course, just the usual relationship between voltage, current, and impedance, applied to the beam in an accelerator.

5.4 Potential-well distortion

In an electron storage ring, a common observation is that the longitudinal profile of a bunch (i.e. the charge per unit length, as a function of position along the bunch) depends on the amount of charge in the bunch (see, for example [20, 21]). In section 3.2.3 we saw that synchrotron radiation effects lead to an equilibrium rms bunch length σ_z given by (3.40)

$$\sigma_z = \frac{\eta_p c}{\omega_s} \sigma_\delta, \tag{5.16}$$

where η_p is the phase slip factor of the lattice (2.77), ω_s is the synchrotron frequency (2.80), and σ_δ is the rms energy spread (3.39). This expression gives the bunch length that we expect in the limit of low bunch charge, but as the bunch charge increases the wake fields that are generated by a bunch lead to an increase in the bunch length. This effect is known as *potential-well distortion* [14]. Usually, potential-well distortion does not have a severe impact on the operation of an electron storage ring, but it can be an interesting effect to study since the behaviour of the bunch can provide some information about the wake fields in a storage ring. If the bunch charge is increased sufficiently, then rather than merely increasing in length, the bunch can become unstable, and this can have an adverse impact on machine performance. Understanding the wake fields is therefore an important aspect of the design and operation of a storage ring.

To estimate the impact of wake fields on the longitudinal dynamics of particles in a storage ring, we can perform an analysis based on the same equations of motion that we used previously (in section 3.2.1), but including an additional term to represent the change in the energy deviation from the effects of wake fields. The equations of motion for the longitudinal co-ordinate z of a particle and the energy deviation δ can be written in the form

$$\frac{dz}{ds} = -\alpha_p \delta, \tag{5.17}$$

$$\frac{d\delta}{ds} = \frac{\omega_s^2}{\alpha_p c^2} z - \frac{e}{E_0 C_0} \int_z^\infty \lambda(z') W_\parallel(z - z') \, dz'. \tag{5.18}$$

Here, α_p is the momentum compaction factor (2.72), E_0 is the reference energy, C_0 is the circumference of the storage ring, and $\lambda(z')$ is the charge per unit length of the bunch, as a function of longitudinal position z' in the bunch. The wake function $W_\parallel(z - z')$ represents the longitudinal wake fields over the entire circumference of the ring. The integral arises from summing the wake fields generated by all particles in the bunch that are ahead of the particle for which the equations of motion are written.

Solving the equations of motion (5.17) and (5.18) is not an easy task; however, we are interested in finding the equilibrium longitudinal distribution, rather than a full solution to the equations of motion. A way of finding the equilibrium longitudinal distribution is provided by the observation that the equations of motion have a conserved quantity[2]:

$$H = -\frac{\alpha_p}{2}\delta^2 - \frac{\omega_s^2}{2\alpha_p c^2}z^2 + \frac{e}{E_0 C_0}\int_0^z dz' \int_{z'}^\infty dz'' \lambda(z'') W_\parallel(z' - z''). \quad (5.19)$$

Since the equilibrium distribution, by definition, remains the same as the bunch moves around the storage ring, it should be possible to express the equilibrium charge distribution as a function of a conserved quantity. Assuming that the charge distribution within a bunch in an electron storage ring is (at low bunch charge) Gaussian, we write the equilibrium distribution $\Psi(z, \delta)$ (the charge per unit area of longitudinal phase space) as

$$\Psi(z, \delta) = \Psi_0 e^H, \quad (5.20)$$

where Ψ_0 is a constant, chosen so that integrating the charge density $\Psi(z, \delta)$ over all values of z and δ gives the total charge in the bunch. If we integrate the equation above for $\Psi(z, \delta)$ just over the energy deviation δ, we obtain

$$\lambda(z) = \lambda_0 \exp\left(-\frac{z^2}{2\sigma_z^2} + \frac{e}{\alpha_p \sigma_\delta^2 E_0 C_0}\int_0^z dz' \int_{z'}^\infty dz'' \lambda(z'') W_\parallel(z' - z'')\right), \quad (5.21)$$

where $\lambda(z)$ is the charge per unit length in the bunch,

$$\lambda(z) = \int_{-\infty}^\infty \Psi(z, \delta)\, d\delta, \quad (5.22)$$

and the constant λ_0 is chosen so that the integral of the charge per unit length over the entire length of the bunch gives the total charge in the bunch. Equation (5.21) is known as the *Haissinski equation* [22]—it is an integral equation for the equilibrium longitudinal charge distribution $\lambda(z)$. If all other quantities are known (including the longitudinal wake function $W_\parallel(z' - z'')$), then it is possible to solve the Haissinski equation to find the longitudinal distribution $\lambda(z)$. Usually, the solution has to be found numerically.

Despite the approximations made in describing the wake fields by means of a wake function, it is often found that (given an appropriate model for the wake function) the Haissinski equation provides a good description of the longitudinal profile of a bunch as a function of the bunch charge in a storage ring. In particular, the Haissinski equation describes the increase in the rms bunch length with increasing bunch charge (see figure 5.4). In physical terms, the wake fields cause a loss of energy of particles in a bunch, and this energy has to be replaced by the RF cavities. As the energy loss increases with increasing bunch charge, the synchronous phase moves closer to the peak of the RF voltage, where the gradient of the RF (rate

---

[2] The conserved quantity H represents the Hamiltonian from which the equations of motion may be derived, using Hamilton's equations; see, for example, [10].

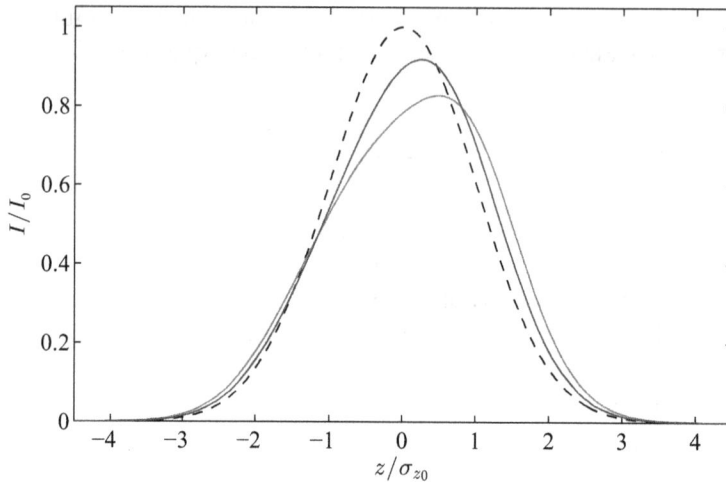

Figure 5.4. Potential-well distortion in an electron storage ring. The horizontal axis indicates the longitudinal position with respect to the reference particle, and the vertical axis indicates the beam current (corresponding to the number of particles per unit length as a function of position along the bunch). At low bunch charge, the bunch has a Gaussian longitudinal profile (dashed black line). At higher bunch charge, wake fields lead to potential-well distortion, resulting in a longitudinal profile that is no longer Gaussian (solid blue line). If the ring is above transition, then the width of the bunch is increased, and the distribution 'tilts' forward. If the bunch charge is increased (solid red line) then the tilt becomes more pronounced.

of change of voltage with time) is smaller. Since the gradient of the RF provides the longitudinal focusing in a storage ring, a smaller slope means less focusing, which results in a longer bunch.

5.5 Microwave instability

Potential-well distortion in an electron storage ring leads to an increase in bunch length, but the energy spread is not significantly changed—as long as the wake fields are not too strong, the energy spread remains close to the value (3.39) expected from the balance between synchrotron radiation damping and quantum excitation. However, if the bunch charge is increased sufficiently, then it is found that above some threshold bunch charge, the energy spread starts to increase as well as the bunch length. This can be explained in terms of an instability, known as the *microwave instability* [23], that is driven by wake fields. In this section, we shall derive an approximate formula for the threshold current at which we expect to see the instability occur.

By their very nature, beam instabilities are phenomena in which the charge distribution continuously evolves in time, without reaching equilibrium. In principle, instabilities may be described by considering the equations of motion of the individual particles within a bunch (or within an entire beam in a storage ring); however, since the number of particles can be very large, this is rarely a practical approach. Instead, we represent a bunch of particles not as a collection of point-like charges, but as a smooth, continuous distribution of electric charge. We can then write an equation that describes how the charge density evolves as a function of time. If we assume that we can neglect dissipative forces (in other words, we assume that the forces acting on the distribution can be derived from suitable potentials) then the appropriate equation is the *Vlasov equation*:

$$\frac{\partial \Psi}{\partial t} + \frac{d\theta}{dt}\frac{\partial \Psi}{\partial \theta} + \frac{d\delta}{dt}\frac{\partial \Psi}{\partial \delta} = 0, \tag{5.23}$$

where $\Psi(\theta, \delta; t)$ is the charge density as a function of the co-ordinate θ (the angular position around the storage ring) and energy deviation δ, at time t. In the case of an electron storage ring, the Vlasov equation describes the evolution of the charge density in a bunch of particles if we neglect synchrotron radiation.

To apply the Vlasov equation to find the evolution of a charge density distribution, we write for the rate of change of the co-ordinate θ

$$\frac{d\theta}{dt} = \frac{2\pi}{T_0}(1 - \eta_{\mathrm{p}}\delta), \tag{5.24}$$

where T_0 is the revolution period for a particle with $\delta = 0$, and η_{p} is the phase slip factor. We also need an explicit expression for the rate of change of the energy deviation. To simplify the analysis, we include only the effects of wake fields, i.e. we neglect the RF cavities and synchrotron radiation. This means that we ignore the synchrotron oscillations of particles in the beam, but if we apply the Vlasov equation only over a time scale that is short compared to a synchrotron period (which may be several hundred turns of the ring) this can be a valid approach. The rate of change of the energy deviation is then given (in terms of the longitudinal impedance $Z_\parallel(\omega)$ of the entire ring) by

$$\frac{d\delta}{dt} = -\frac{e}{E_0 T_0}\int_{-\infty}^{\infty}\frac{d\omega}{2\pi}e^{-i\omega t}\tilde{I}(\omega)Z_\parallel(\omega), \tag{5.25}$$

where e is the magnitude of the charge on a single particle (electron, or positron) in the beam, and $\tilde{I}(\omega)$ is the beam current spectrum. The expressions for the rate of change of co-ordinate θ (5.24) and energy deviation δ (5.25) are substituted into the Vlasov equation (5.23); the solution of the partial differential equation that we obtain in this way gives the evolution of the phase space density $\Psi(\theta, \delta; t)$ as a function of time t, for a given initial distribution $\Psi(\theta, \delta; 0)$.

In practice, solution of the Vlasov equation usually needs to be done numerically. However, we can obtain some useful results by assuming an approximate solution of the form

$$\Psi(\theta, \delta; t) = \Psi_0(\delta) + \Delta\Psi e^{i(n\theta - \omega_n t)}, \tag{5.26}$$

where $\Psi_0(\delta)$ is an assumed constant energy distribution, and $\Delta\Psi$ is the amplitude of a modulation in the density of the particles as a function of position around the ring. Note that the modulation takes the form of a wave travelling around the ring, with wavelength C_0/n and angular frequency ω_n. By substituting the assumed solution (5.26) into the Vlasov equation (5.23), and integrating over the energy deviation (to eliminate the unknown amplitude $\Delta\Psi$) we obtain an integral equation for the oscillation frequency ω_n of the density modulation[3]:

$$1 = -iZ_{\parallel}(\omega_n)\frac{eI_0}{E_0 T_0} \int_{-\infty}^{\infty} \frac{\partial\Psi_0/\partial\delta}{n\omega - \omega_n}d\delta, \tag{5.27}$$

where I_0 is the average beam current, and $\omega = d\theta / dt$ is the angular revolution frequency of a particle with energy deviation δ. The above equation (5.27) is a *dispersion relation*; it relates the frequency of the wave (representing the density modulation) to its wavelength. Given the impedance of the ring $Z_{\parallel}(\omega)$, the energy spread $\Psi_0(\delta)$ and the beam current I_0, we can solve the dispersion relation (5.27) to find the oscillation frequency ω_n of the density modulation as a function of the wavelength C_0 / n. If the frequency is a real number, then the modulation will simply propagate as a wave with constant amplitude. However, if the frequency is a complex number, then depending on the sign of the imaginary part, the amplitude of the modulation will either damp or grow exponentially. In the latter case, an initially small modulation can rapidly become very large, indicating an instability in the beam.

5.5.1 Microwave instability in a 'cold' beam

As an example, consider a beam with zero energy spread, i.e. with $\Psi(\delta) = 0$ for all δ except $\delta = 0$. Such a beam is sometimes called a 'cold' beam. The dispersion relation in this case can be solved to give

$$\frac{\omega_n}{n\omega_0} = 1 \pm \sqrt{i\frac{Z_{\parallel}(\omega_n)}{n}\frac{eI_0\eta_p}{2\pi E_0}}. \tag{5.28}$$

We see that in this case there will (unless the impedance is a purely imaginary number) always be a solution for the modulation frequency ω_n with a positive imaginary part. Since a beam will inevitably never have a perfectly uniform distribution around a storage ring, a cold beam will always be unstable.

5.5.2 Energy spread and beam stability: Landau damping

A more realistic example is the case where the beam has a Gaussian energy spread:

$$\Psi_0 = \frac{e^{-\delta^2/2\sigma_\delta^2}}{\sqrt{2\pi}\,\sigma_\delta}. \tag{5.29}$$

In this case, it is not possible to write an analytical solution for ω_n, but it can be shown that for an impedance such that $Z_{\parallel}(n\omega_0)/n$ is approximately constant, that the beam will be stable (the modulation frequency ω_n will have a negative imaginary part) if the beam current is below the instability threshold:

---

[3] The details of the calculation are given, for example, in [10, 14].

$$I_{th} = \frac{\pi^2 \sqrt{2\pi}}{3} |\eta_p| \sigma_\delta^2 \frac{E_0/e}{|Z_\parallel(n\omega_0)n|}. \tag{5.30}$$

It is only if the current is above I_{th} that the beam becomes unstable. This is in contrast to a cold beam, which was unstable for any size of beam current. The reason that the energy spread can stabilise the beam is that the revolution frequency of the particles depends (through the phase slip factor η_p) on the energy deviation. Particles with different energies will move around the ring at different rates, and this tends to smooth out any modulation that appears in the particle density. The suppression of an instability in this way is known (by analogy with a similar effect in plasma physics) as *Landau damping* [24, 25]. However, if the current is large enough then the enhancement of the modulation amplitude by wake fields outweighs its suppression by Landau damping, and the modulation amplitude is able to grow exponentially (indicating a beam instability).

Given the numerous simplifications and approximations we have made in the analysis, we are not able to determine with any precision how a density modulation will develop in any given situation—all we can say is whether the beam is likely to be stable or unstable, and even then there can be considerable uncertainty. We also need to be careful about applying the results to an electron storage ring, since we have neglected the effects of RF cavities and synchrotron radiation. Strictly speaking, our analysis applies to the case that there is an approximately uniform distribution of charge around the ring. However, it turns out that in practice, the wavelength of the modulation in the case of an instability is often much shorter than the bunch length; one indication of this is that the instability can be accompanied by the emission of electromagnetic radiation with wavelengths of 1 mm or less, associated with a large density modulation in an individual bunch. Because of this radiation, the instability is usually called the *microwave instability*. Given the short wavelength of the modulation, it is often assumed that the analysis can be applied to bunched beams, rather than just to beams with approximately uniform density around the entire ring, in which case we simply substitute the peak current in the bunch for the average current in the ring. The stability condition (5.30) is then usually expressed in terms of the impedance rather than the current:

$$\left| \frac{Z_\parallel(n\omega_0)}{n} \right| < \sqrt{\frac{\pi}{2}} Z_0 \frac{\gamma |\eta_p| \sigma_\delta^2 \sigma_z}{r_e N_b}, \tag{5.31}$$

where Z_0 is the impedance of free space, σ_z is the rms bunch length, r_e is the classical radius of the electron, and N_b is the number of electrons in a single bunch. The stability condition (5.31) is known as the *Keil–Schnell–Boussard criterion* [23, 26].

In the analysis in this section, we have considered only the longitudinal motion of the beam. Of course, wake fields can also affect the transverse motion, leading to beam instabilities with a range of characteristics. Transverse instabilities observed in electron storage rings include the *head–tail instability*, and the *transverse mode-coupling instability*. In principle, the Vlasov equation can be extended to include transverse motion as well as longitudinal motion; however, different analysis methods can also be used. For further information, the reader is referred to other texts, for example [14].

5.6 Coupled-bunch instabilities

The effects that we considered in the previous sections, potential-well distortion and the microwave instability, result from wake fields acting over the length of a single bunch in an electron storage ring, which is typically a few millimetres. We assumed that we could neglect any effects from (long-range) wake fields acting over the distance between individual bunches, which could be many centimetres or some metres. However, long-range wake fields can affect the stability of a beam in a storage ring, and in this section we shall consider how coupled-bunch instabilities develop, how they may affect machine performance, and how they may be mitigated.

We shall base our analysis on a simple model of a beam, in which each bunch is represented as a point-like object, with total charge eN_b and mass mN_b, where e and m are the magnitude of the charge and the mass of the electron, respectively, and N_b is the bunch population. Since we ignore any internal structure in a bunch, this approach is limited to *coherent* betatron and synchrotron oscillations, where the entire bunch oscillates transversely or longitudinally as it moves around the storage ring. Here, we shall consider only betatron oscillations; however, synchrotron oscillations may be treated in much the same way. To simplify the model further, we shall assume that the betatron oscillations can be characterised by an angular frequency ω_β that is the same at all points around the ring. In effect, this neglects any variation in the Courant–Snyder beta function. With these simplifications, the equation of motion for a bunch performing vertical betatron oscillations in the absence of wake fields is

$$\frac{d^2y}{dt^2} + \omega_\beta^2 y = 0, \qquad (5.32)$$

where y is the vertical co-ordinate of the bunch centroid (i.e. the centre of mass of the bunch).

We can include the effects of wake fields by adding a driving term on the right-hand side of the equation of motion (5.32). To start with, let us assume that there is only a single bunch in the storage ring. At each point in the storage ring, the bunch will generate a wake field that will act back on the bunch on later turns. In terms of the wake function (5.7) that describes the change in momentum of a particle (or, in this case, an entire bunch of particles) resulting from the wake field, the equation of motion for the bunch centroid can be written

$$\frac{d^2y}{dt^2} + \omega_\beta^2 y = -\frac{N_b e^2 c^2}{E_0 C_0} \sum_{k=1}^{\infty} W_\perp(-kC_0)\, y(t - kC_0/c), \qquad (5.33)$$

where C_0 is the ring circumference, and the index k refers to the number of turns around the ring that the bunch has performed since it generated the wake field at its present point. Since the summation over k extends to infinity, there is an assumption that the bunch has been in the ring for an infinite length of time; in practice, since the wake fields will decay over time, the summation may be truncated at some finite value of k for which the wake function becomes negligible.

To include the effects of multiple bunches in the ring, we need to add another summation, to take account of the wake fields generated by each bunch. We write the centroid of the mth bunch at time t as $y_m(t)$. Then, if there are M bunches in the storage ring we can write the equation of motion for the nth bunch as

$$\frac{d^2 y_n}{dt^2} + \omega_\beta^2 y_n = - \frac{N_b e^2 c^2}{E_0 C_0} \sum_{k=1}^{\infty} \sum_{m=1}^{M} W_\perp \left(-kC_0 - \frac{m-n}{M} C_0 \right)$$
$$y_m \left(t - k\frac{C_0}{c} - \frac{m-n}{M}\frac{C_0}{c} \right). \tag{5.34}$$

Note that we assume that the bunches are equally spaced around the storage ring, so that the distance between the mth and the nth bunches is $(m-n)C_0/M$. We also assume that all bunches have the same population N_b. To understand the dynamics of the beam in the presence of wake fields, we need to find a solution to the equation of motion (5.34). We shall assume a solution of the form

$$y_n^\mu(t) = A \exp\left(2\pi i \frac{\mu n}{M}\right) e^{-i\Omega_\mu t}, \tag{5.35}$$

where μ is an index describing different patterns of variation of the centroid co-ordinate y_m from bunch-to-bunch, at a given time t. For example, for $\mu = 0$, all bunches have the same centroid co-ordinate (at any time). For $\mu > 0$ the bunch centroids form a wave around the ring with wavelength C_0/μ. Each pattern of bunch centroids corresponds to a different *mode*; any arbitrary set of bunch centroids can be constructed by adding together different modes with appropriate amplitudes. An example, the modes in a storage ring with six bunches are shown in figure 5.5. Since different modes will generate different wake fields within the ring, the oscillation frequency of bunches moving around the ring will depend on the mode—we write the oscillation frequency Ω_μ. We can assume that, if the wake fields are not too strong, the mode oscillation frequencies will all be close to the betatron frequency, i.e. $\Omega_\mu \approx \omega_\beta$.

We are mainly interested in whether, for a given set of parameters and a given wake function, the beam will be stable or not. Information on the beam stability can be obtained from the mode frequencies Ω_μ. If a frequency Ω_μ is a real number, then bunches in that mode will oscillate at that frequency with constant amplitude as they move around the ring. However, if Ω_μ has a positive imaginary part, then the amplitude of the bunch oscillations will increase exponentially, with growth rate given by the imaginary part. Since an arbitrary pattern of bunch centroid co-ordinates will, in general, contain some component from every possible mode, it is only necessary for one mode to have a positive growth rate for the beam in the storage ring to be unstable.

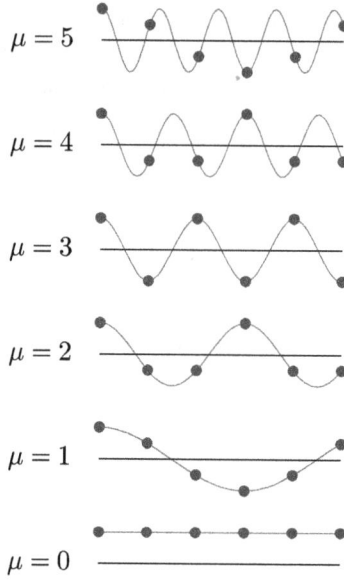

$\mu = 5$

$\mu = 4$

$\mu = 3$

$\mu = 2$

$\mu = 1$

$\mu = 0$

Figure 5.5. Multibunch modes in a storage ring. The number of distinct modes (six in this example) is the same as the number of bunches in the ring. In a given mode, the transverse positions of successive bunches (blue circles) at a particular moment in time form a 'wave' around the ring—each mode corresponds to a different wavelength. Any pattern of bunch positions can be constructed by superposing the various modes with appropriate amplitudes.

Our goal now is to find an expression for the mode frequencies Ω_μ; this can be done by substituting the assumed solution (5.35) into the equation of motion (5.34). The result can most easily be expressed in terms of the impedance $Z_\perp(\omega)$ rather than the wake function $W_\perp(-\Delta z)$; we find (with some approximations) [10, 14]

$$\Omega_\mu \approx \omega_\beta - i\frac{I_b ec}{4\pi\nu_y E_0} \sum_{p=-\infty}^{\infty} Z_\perp((pM + \mu)\omega_0 + \omega_\beta), \tag{5.36}$$

where ν_y is the vertical betatron tune, $I_b = MN_b ec/C_0$ is the beam current, and ω_0 is the angular revolution frequency of a bunch moving around the ring. Given a set of parameters for the storage ring and the beam, and a wake function (or impedance) we can apply the above expression (5.36) to determine the growth rates of the different modes.

As an example, let us consider the wake function (5.9) for the resistive-wall wake field. The impedance in this case is given by [14, 18]:

$$Z_\perp(\omega) = (1 - i \, \mathrm{sgn}(\omega))\frac{C_0}{\omega r^3}\sqrt{\frac{Z_0 c}{4\pi}\frac{2|\omega|}{\pi\sigma}}, \tag{5.37}$$

where the beam pipe has conductivity σ and uniform circular cross-section with radius r, and Z_0 is the impedance of free space. Note that the impedance becomes large at low frequencies. This means that the beam may be strongly affected by the wake fields if it is oscillating in a mode such that the co-ordinate of the bunch centroids observed at a fixed point in the storage ring change slowly over time, i.e. such that each successive bunch arrives at any given point in the ring with approximately the same centroid co-ordinate y_m. In fact, it is found from (5.36) that in the case of resistive-wall wake fields, the mode with the highest growth rate has mode number μ that minimises the value of $pM + \mu + \nu_y$ for an integer p. The growth rate of this mode is given by

$$\frac{1}{\tau_{\min}} = \frac{I_b ec C_0}{4\pi^2 \nu_y E_0 r^3} \sqrt{\frac{Z_0 c}{2\sigma\omega_0\left(1 - \text{frac}(\nu_y)\right)}}, \qquad (5.38)$$

where $\text{frac}(\nu_y)$ is the fractional part of the betatron tune. The growth or damping rates of the modes driven by resistive-wall wake fields in a storage ring with 50 bunches are shown in figure 5.6.

In an electron storage ring for a third-generation synchrotron light source, resistive-wall wake fields can drive coupled-bunch modes with growth times of hundreds or tens of turns. In practice, beams may be stable at low beam current because synchrotron radiation and decoherence provide natural damping mechanisms that can be strong enough to suppress the instability. However, as more current is injected into the ring, the wake fields eventually become strong enough for the beam to be unstable; the oscillations of individual bunches will then grow in amplitude until some particles are lost from the beam, reducing the current and

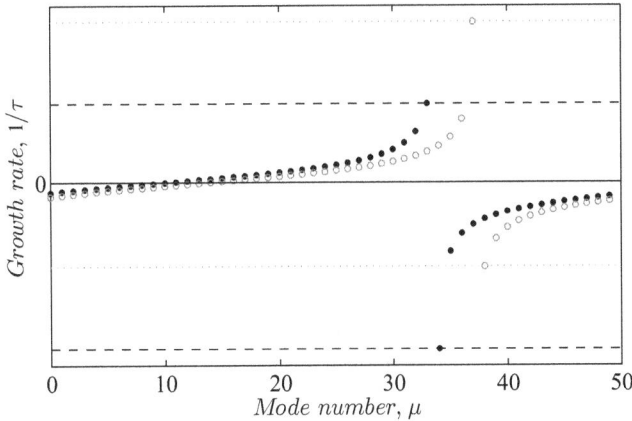

Figure 5.6. Growth and damping rates of different modes in a storage ring with resistive-wall wake fields. In this case, there are 50 bunches in the ring. Positive growth rates indicate unstable (antidamped) modes, and negative growth rates indicate stable (damped) modes. Solid black points show the growth or damping rates of different modes in the case that the betatron tune is 16.2—the integer part of the tune determines which mode has the fastest growth or damping rate. The fact that the fastest mode (number 34) in this case is a damped mode follows from the fact that the fractional part of the tune is less than 0.5. The red circles show the growth or damping rates in the case that the betatron tune is 12.8. The fastest mode is now mode number 37, and because the fractional part of the tune is greater than 0.5, this mode is an antidamped mode. The horizontal black dashed and red dotted lines show the maximum growth and damping rates for the cases with tune 16.2 and 12.8, respectively. The variation of the growth or damping rate with mode number is characteristic of the wake fields present in the storage ring (in this case, resistive-wall wake fields).

restoring beam stability. This means that coupled-bunch instabilities can appear as a current limit in machine operation, preventing the injection of beam currents larger than the threshold for instability set by the wake fields.

To achieve the performance specifications of the storage ring, however, it may be necessary to operate with beam currents significantly larger than the instability threshold. In that case, it is possible to use a *bunch-by-bunch feedback system* to maintain beam stability [27–29]. In its simplest form, a bunch-by-bunch feedback system consists of a beam position monitor (BPM) capable of detecting the co-ordinates of each bunch on each turn, a high-power amplifier, and a fast kicker that can deflect each bunch in the beam as it passes (see figure 5.7). In principle, the signal from the BPM can be amplified and fed to the kicker so that any bunch with some transverse offset from the closed orbit is deflected back towards the closed orbit. Usually, the correction to each bunch trajectory needs to be done gradually over some number of turns because of limitations in the technology of the feedback system. For example, causality can prevent the signal detected on one turn being fed back to a given bunch on the same turn, because the bunches are travelling at close to the speed of light. Although this limitation may be overcome, in principle, by arranging for the feedback signal to take a shorter path from the BPM to the kicker than the beam (across a diameter of the storage ring, for example), amplification and processing of the signal inevitably introduces some delay in the system. Nevertheless, bunch-by-bunch feedback systems can be constructed and operated to provide damping of coherent betatron (and synchrotron) oscillations, with damping times of some tens of turns [30, 31]. This is usually sufficient to maintain beam stability in the parameter regimes needed by modern light sources and colliders. It should be mentioned, however, that the first mitigation to consider is normally to minimise the wake fields as far as possible. For resistive-wall wake fields, this means (for example)

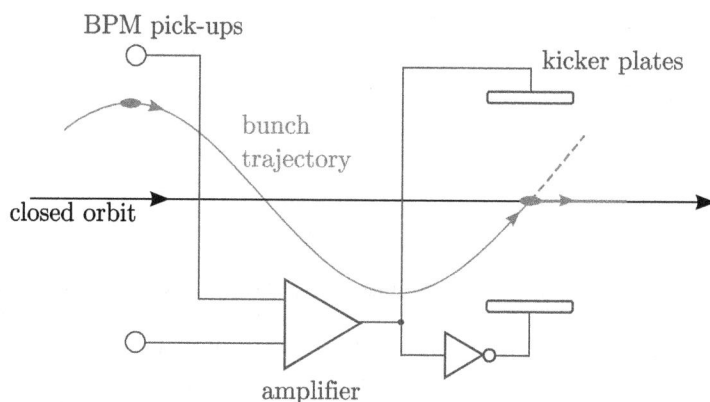

Figure 5.7. Principle of a fast feedback system for damping coupled-bunch instabilities. The signal from a bunch passing through a BPM is amplified and used to control the deflection applied by a kicker. If the phase advance from the BPM to the kicker is an odd multiple of $\pi/2$, the bunch is deflected towards the closed orbit. Because of practical limitations from the strength of the kicker, the speed of the amplifier and the level of noise in the system, the correction cannot be achieved in a single pass, but must be applied over several turns of the ring.

having a beam pipe with a large diameter, and using a material with a good electrical conductivity, such as aluminium or copper. However, long-range wake fields can also come from other sources, such as cavities and transitions in the vacuum chamber, and each section of the vacuum chamber needs to be carefully designed to minimise the wake fields that will be generated by the beam in that section.

References

[1] Roqué X 1991 Møller scattering: a neglected application of early quantum electrodynamics *Arch. Hist. Exact Sci.* **44** 197–264

[2] Piwinski A 1974 Intra-beam scattering *Proc. 9th Int. Conf. on High Energy Accelerators (Stanford, CA, USA)* pp 405–8.

[3] Bjorken J D and Mtingwa S K 1983 Intrabeam scattering *Part. Accel.* **13** 115–43

[4] Kubo K and Oide K 2001 Intrabeam scattering in electron storage rings *Phys. Rev. Spec. Top. - Accel. Beams* **4** 124401

[5] Kubo K *et al* 2002 Extremely low vertical emittance beam in the Accelerator Test Facility at KEK *Phys. Rev. Lett.* **88** 194801

[6] Steier C, Robin D, Wolski A, Portmann G and Safranek J 2003 Coupling correction and beam dynamics at ultralow vertical emittance in the ALS *Proc. 2003 Particle Accelerator Conference (Portland, OR, USA)* pp 3213–5.

[7] Bernardini C, Corazza G F, Di Giugno G, Ghigo G, Haissinski J, Marin P, Querzoli R and Touschek B 1946 Lifetime and beam size in a storage ring *Phys. Rev. Lett.* **10** 407–9

[8] Wiedemann H 2015 *Particle Accelerator Physics* 4th edn (New York: Springer)

[9] Conte M and Mackay W W 2008 *An Introduction to the Physics of Particle Accelerators* 2nd edn (Singapore: World Scientific)

[10] Wolski A 2014 *Beam Dynamics in High Energy Particle Accelerators* (London, UK: Imperial College Press)

[11] Franchetti G 2017 Space charge in circular machines *Proc. CERN Accelerator School, Intensity Limitations in Particle Beams (Geneva, Switzerland)* ed H Werner number CERN-2017-006-SP pp 353–90

[12] Zimmermann F 2013 Ion trapping, beam-ion instabilities, and dust *Handbook of Accelerator Physics and Engineering* 2nd edn ed A Wu Chao, K H Mess, M Tigner and F Zimmermann (Singapore: World Scientific) pp 159–63

[13] Nagaoka R 2017 Ions *Proc. CERN Accelerator School, Intensity Limitations in Particle Beams (Geneva, Switzerland)* ed H Werner number CERN-2017-006-SP pp 519–56

[14] Wu Chao A 1993 *Physics of Collective Beam Instabilities in High Energy Accelerators* (New York: Wiley)

[15] Dohlus M and Wanzenberg R 2017 An introduction to wake fields and impedances ed W Herr *Proceedings of the CERN Accelerator School, Intensity Limitations in Particle Beams, number CERN-2017-006-SP,* (Geneva, Switzerland) pp 15–41

[16] Gluckstern R L and Kurennoy S S 2013 Impedance calculation, frequency domain *Handbook of Accelerator Physics and Engineering* 2nd edn ed A Wu Chao, K H Mess, M Tigner and F Zimmermann (Singapore: World Scientific) pp 243–8

[17] Gjonaj E and Weiland T 2013 Impedance calculation, time domain *Handbook of Accelerator Physics and Engineering* 2nd edn ed A Wu Chao, K H Mess, M Tigner and F Zimmermann (Singapore: World Scientific) pp 248–52

[18] Ng K Y and Bane K 2013 Explicit expressions of impedances and wake functions *Handbook of Accelerator Physics and Engineering* 2nd edn ed A Wu Chao, K H Mess, M Tigner and F Zimmermann (Singapore: World Scientific) pp 252–62

[19] Suzuki T 2013 Definitions and properties of impedances and wake functions *Handbook of Accelerator Physics and Engineering* 2nd edn ed A Wu Chao, K H Mess, M Tigner and F Zimmermann (Singapore: World Scientific) pp 242–3

[20] Lumpkin A H, Chang B X and Chae Y C 1997 Observations of bunch-lengthening effects in the APS 7-GeV storage ring *Nucl. Instrum. Methods Phys. Res. Sect.* A **393** 50–4

[21] Cai Y, Flanagan J, Fukuma H, Funakoshi Y, Ieiri T, Ohmi K, Oide K and Suetsugu Y 2009 Potential-well distortion, microwave instability, and their effects with colliding beams at KEKB *Phys. Rev. Spec. Top. - Accel. Beams* **12** 061002

[22] Haissinski J 1973 Exact longitudinal equilibrium distribution of stored electrons in the presence of self-fields *Il Nuovo Cimento* B **18** 72–82

[23] Keil Eberhard and Schnell W 1969 Concerning longitudinal instability in the ISR *Technical Report CERN–ISR–TH–RF/69–48* (Geneva, Switzerland: CERN)

[24] Landau Lev 1946 On the vibration of the electronic plasma *J. Phys. (USSR)* **10** pp 25–34

[25] Werner H 2017 Introduction to Landau damping *Proc. CERN Accelerator School, Intensity Limitations in Particle Beams (Geneva, Switzerland)* ed W Herr number CERN-2017-006-SP pp 137–64

[26] Boussard Daniel 1975 Observation of microwave longitudinal instabilities in the CPS *Technical Report LABII/RF/Int./75–2* (Geneva, Switzerland: CERN)

[27] Lonza M 2009 Multi-bunch feedback systems *Proc. CERN Accelerator School, Synchrotron Radiation and Free Electron Lasers (Dourdan, France)*, number CERN–2009–005, pp 467–511

[28] Lonza M 2017 Multi-bunch feedback systems *Proc. CERN Accelerator School, Intensity Limitations in Particle Beams (Geneva, Switzerland)* ed W Herr number CERN-2017-006-SP pp 471–514

[29] Fox J D 2013 Feedback to control coupled-bunch instabilities *Handbook of Accelerator Physics and Engineering* 2nd edn ed A W Chao, K H Mess, M Tigner and F Zimmermann (Singapore: World Scientific) pp 628–36

[30] Fox J D, Larsen R, Prabhakar S, Teytelman D, Young A, Drago A, Serio M, Barry W and Stover G 1999 Multi-bunch instability diagnostics via digital feedback systems at PEP-II, DAΦNE, ALS and SPEAR *Proc. 1999 Particle Accelerator Conference (New York)* pp 636–40

[31] Fox J D, Mastorides T, Rivetta C, Van Winkle D and Teytelman D 2008 Lessons learned from PEP-II LLRF and longitudinal feedback. *Proc. Eleventh European Particle Accelerator Conference (Genoa, Italy)* pp 1953–1955

Introduction to Beam Dynamics in High-Energy Electron Storage Rings

Andrzej Wolski

Chapter 6

Further topics

6.1 Advanced tools for beam dynamics

Many of the beam dynamics phenomena in a storage ring can be described and analysed mathematically using relatively straightforward techniques. For example, the motion of a particle through an electromagnetic field can be calculated using Newton's laws of motion (taking into account relativistic effects, as necessary), with the force $\vec{F}$ given by the Lorentz formula

$$\vec{F} = q(\vec{E} + \vec{v} \times \vec{B}), \tag{6.1}$$

where q is the electric charge of the particle, $\vec{E}$ and $\vec{B}$ are the electric and magnetic fields, and $\vec{v}$ is the velocity of the particle. The linear optics outlined in chapter 2 are developed by expressing the solutions to the equations of motion for individual components in the form of matrices (transfer matrices), and then simply multiplying the matrices to describe the motion of particles through a beamline consisting of any given sequence of components. For the design, commissioning, and operation of a basic storage ring, little more than this is needed.

Modern storage rings, however, frequently have demanding performance specifications in terms of beam quality, stability, and intensity. In the parameter regimes that then need to be considered, a large number of complex and often rather subtle phenomena can have a significant impact on machine design and performance. To assist the analysis of some of these phenomena and to provide the understanding needed to control or avoid potentially damaging effects, a variety of sophisticated mathematical and computational tools have been developed. Here, we outline just a few of them.

In the case of nonlinear dynamics (arising, for example, in the context of dynamic aperture and energy acceptance) many of the most powerful mathematical techniques are based on Hamiltonian, rather than Newtonian, mechanics [1]. The

Hamiltonian formalism has the same physical content as Newtonian mechanics, but instead of specifying a force that determines (through Newton's second law of motion) the dynamics of a given system, the dynamics are determined by a function, the *Hamiltonian*, from which the equations of motion can be derived using Hamilton's equations. In many (but not all) cases, the Hamiltonian corresponds to the energy of the system. The main benefit of using the Hamiltonian formalism is that the equations of motion are more readily applied to co-ordinate systems that are more complicated than a simple Cartesian co-ordinate system, and can be written so that the solutions naturally appear as functions of some variable other than the time. This is of benefit for the study of beam dynamics in an accelerator, where (as discussed in section 2.1) it is convenient to use a co-ordinate system based on a curved reference trajectory for the transverse co-ordinates x and y, the longitudinal co-ordinate z specifies the time at which a particle arrives at a particular location relative to a given reference particle, and we wish to express the co-ordinates as functions of the distance s along the reference trajectory.

Another benefit of Hamiltonian mechanics that is often of value for accelerator beam dynamics is that the formalism can make conserved quantities, when they appear, more explicit or apparent. For example, neglecting effects such as synchrotron radiation, the betatron action J_x (see section 2.2.1) is constant for a particle moving along a beamline. In the Hamiltonian formalism, the action is a momentum associated with a co-ordinate (the betatron phase, ϕ_x). The fact that the betatron action is constant under certain conditions is then a form of conservation of momentum, and is immediately apparent from Hamilton's equations. Hamilton's equations describe the motion of a particle in terms of *action-angle variables* (J_x, ϕ_x) [1] as readily as in terms of the 'Cartesian' variables (x, p_x). The drawback is that one needs to obtain an expression for the Hamiltonian in terms of the particular set of variables being used—but there are established procedures for doing this, and (in principle) it only needs to be done once[1]. Making a suitable choice of variables (co-ordinates and momenta) for describing the dynamics of a given system can greatly simplify the analysis of the system: the Hamiltonian formalism provides convenient and systematic procedures for converting between different sets of variables.

Hamiltonian mechanics also provides the basis for some powerful techniques for finding solutions to the equations of motion. In many cases of interest in beam dynamics, exact solutions to the equations of motion cannot be written down in a closed algebraic form (i.e. in an expression consisting of a finite number of terms). This is the case, for example, for a particle moving through a sextupole magnet (or any higher-order multipole magnet). In such cases, it can be useful to have techniques for constructing approximate solutions that can provide, in principle, any desired degree of accuracy, and that satisfy certain constraints regarding conserved quantities (such as energy and momentum). One technique that can be of significant value in the study of nonlinear dynamics makes use of *Lie transformations* [3, 4]. The application of Lie transformations in this context is based on

---

[1] For derivation of the Hamiltonian in terms of the variables used in this text, see [2].

the fact that the evolution of any function f of the dynamical variables (the co-ordinates and the momenta) can be written as

$$\frac{df}{ds} = -:H: f, \tag{6.2}$$

where H is the Hamiltonian, and s is the independent variable (corresponding, in the case of an accelerator beamline, to the distance along the reference trajectory). The notation $:H:$ indicates a *Lie operator* constructed from the function H, and defined, for motion in just one degree of freedom, by

$$:H := \frac{\partial H}{\partial x}\frac{\partial}{\partial p_x} - \frac{\partial H}{\partial p_x}\frac{\partial}{\partial x}. \tag{6.3}$$

This definition is readily extended to motion in more than one degree of freedom by adding corresponding terms in the derivatives with respect to the additional variables. Given the 'equation of motion' (6.2) expressed in terms of a differential operator, the solution is expressed in terms of a Lie transform, which is the exponential of a Lie operator

$$f(s_1) = e^{-\Delta s:H:}f \big|_{s=s_0}, \tag{6.4}$$

where the exponential is defined in terms of its usual power series expansion, $\Delta s = s_1 - s_0$, and the expression on the right-hand side is to be evaluated (after application of the Lie transform) at $s = s_0$. The advantage of using Lie transformations is that they provide a means of writing the solution to the equation of motion in terms of a series of derivatives. In effect, a problem of integration (which, in general, is difficult to solve) is turned into a problem of differentiation (which can be solved by applying systematically a set of known rules). Given a Hamiltonian of a particular form for a particle in an electromagnetic field, Lie transforms provide a method for construct-ing an expression, in the form of a series, describing the motion of the particle through the field. In many cases, the exact solution can only be expressed using an infinite number of terms; however, a practical solution achieving a specified degree of accuracy can often be obtained by truncating the series after some number of terms (or up to a given power in an appropriate 'small' parameter). The truncation needs to be done with care; although the exact solution will obey certain conservation laws, these laws will not be satisfied, in general, if the infinite series is truncated.

One conservation law that is often of importance in accelerator beam dynamics is associated with a property of Hamiltonian systems known as *symplecticity*. This ensures, for example, that the beam emittances are constants of the motion (if radiation and collective effects are neglected), and is connected with *Liouville's theorem*, which states that for any Hamiltonian system the density of points in phase space is conserved as the system evolves [1, 2, 5]. As explained above, if Lie transforms are applied to solve the equations of motion for particles in accelerators, the result is often obtained as a series that must be truncated at some point; the truncation usually means that the solution is not symplectic, and can then lead to non-physical growth or damping of the emittances. However, techniques have been developed for making approximations to the Hamiltonian that allow solutions to the equations of motion in terms of *finite* series [6, 7]. Such techniques are generally known as *explicit symplectic integrators* and can be of significant importance in some aspects of beam dynamics.

Analytical tools, such as those provided by Lie transformations, can be of great value in accelerator physics; for example, in the construction of transfer maps, and the analysis of nonlinear dynamics. However, in most practical cases, accelerator beamlines rapidly become too complicated for analytical techniques to provide the most appropriate solution. For this reason, numerical methods, usually implemented in computer codes, are often used in real-world situations. The range of tasks and phenomena that need to be addressed (including linear optics design, nonlinear optimisation, modelling of collective instabilities) is rather large, and this has led over the years to the development of a considerable number of computer codes. Some codes are quite specific in the type of problem or variety of machine to which they can be applied, others are more general (such as MAD-X [8], Bmad [9], and elegant [10]). Another distinction that can be made, is between codes designed to work with a machine description and set of commands provided in a customised input file, and those designed as 'libraries' of functions that can be called from a simulation developed in a standard software development environment (e.g. using C++ or Python). The choice of code to use in a particular instance is usually based primarily on the capabilities required, but the way in which the code is used is also an important criterion.

The ways in which the various aspects of accelerator physics can be implemented in a code also vary widely. The basic function of an optics code, for example, is to track particles through a beamline and calculate quantities such as the Courant–Snyder parameters and the tunes. Most codes will track particles simply by applying the appropriate transfer matrices (or transfer maps, including higher-order terms) to a set of initial conditions given as specific numerical values. However, codes have also been developed that can manipulate transfer maps in algebraic form, using techniques such as *differential algebra* [11, 12]; examples include COSY INFINITY [13] and PTC [8, 14, 15]. These codes allow (amongst other things) the construction of a transfer map in the form of an algebraic expression, for a section of beamline consisting of a large number of components. This means that the map can be applied easily to any given (numerical) initial conditions, allowing fast tracking of particles through the beamline. It is also possible, given a map in appropriate form for a storage ring, to perform some analysis to determine the strengths of different resonances and to estimate their impact on the dynamic aperture (see, for example, [16]).

6.2 Some other phenomena

Although we have aimed in the preceding chapters to focus on topics that will be of importance for many cases, design and operation of an electron storage ring usually requires attention to be given to a range of further issues. Here, we mention (briefly) just three of the phenomena that may be important in particular situations: spin dynamics, beam-beam effects, and electron cloud effects.

6.2.1 Spin dynamics

In addition to the angular momentum that they may have as a result of their motion through space, electrons (and positrons) carry an intrinsic angular momentum, known as *spin*. The spin of a particle is a property arising from quantum mechanics and is fixed for a particular type of particle. The spin of an electron or positron is $\frac{1}{2}\hbar$, where $\hbar$ is Planck's constant divided by 2π. Spin is a vector quantity, having both magnitude and direction—the direction of the spin is the direction of the axis of 'rotation' (although it should be remembered that it is not strictly accurate to regard elementary particles as rotating physically in space). Although it is not possible in an accelerator to achieve a situation in which the spin of every particle within a bunch has the same orientation, under some circumstances there can be a predominance of particle spins in one particular direction. The extent to which the spins of a bunch of particles are aligned is quantified by the *spin polarisation* of the bunch.

The main significance of spin polarisation for accelerator beam dynamics is that the interactions between particles often have a dependence on the spins of the particles involved. For example, when two particles collide at high energy producing new particles, the kinds of particles that can be produced (and the probability for their production) depends, amongst other things, on the orientation of the spins of the colliding particles. This means that in a collider, it can be important to be able to control the spins of the particles within the colliding beams. The rate of Touschek scattering is also sensitive to the orientations of the spins of the particles involved in the scattering process—particles with parallel spins have a lower probability of scattering than particles with antiparallel spins [17]. This means that the Touschek lifetime in a storage ring will be slightly longer if the beam has a high degree of spin polarisation [18].

In an electron storage ring, synchrotron radiation naturally causes a beam that starts with zero polarisation (i.e. with particle spins uniformly distributed in all possible directions) to develop some degree of spin polarisation. This is because radiation also carries angular momentum—photons have intrinsic (spin) angular momentum $\hbar$. As a result, when an electron emits a photon, it is possible that the direction of the spin of the electron will be reversed. This is more likely to happen for an electron in a dipole magnet if the spin of the electron is initially aligned opposite to the direction of the magnetic field, since the spin of the electron is associated with its magnetic dipole moment, and the magnetic energy of the electron is lower if its dipole moment is parallel to an external magnetic field. The consequence of this is that over a period of typically several minutes or hours, a beam in an electron storage ring can achieve a high degree of polarisation with the spins of the electrons predominantly oriented parallel to the field in the dipole magnets. The polarisation of a beam in this way is known as the *Sokolov–Ternov effect* [19].

The equilibrium level of spin polarisation in an electron storage ring is determined not just by the rate of spin reversal resulting from photon emission, but also by the dynamics of the spins of individual particles moving around the storage ring. The fact that the spin is associated with a magnetic dipole moment means that when an electron is in a magnetic field, its spin will precess around the direction of the magnetic field; that is, the component of the spin perpendicular to the magnetic field will steadily rotate about an axis parallel to the field. For a particle in a storage ring, the rate of precession in a given electromagnetic field is expressed by the *Thomas–Bargmann–Michel–Telegdi equation* (usually known as the 'Thomas–BMT equation') [20, 21]. If the spin of a particle is parallel to the field of the dipole magnets in a storage ring, then the orientation of the spin will remain constant as the particle moves around the ring (although the field in the quadrupole magnets and other multipole magnets will affect the spin if the particle passes through these magnets with some displacement from the magnetic axis). If the spin of the particle has some component perpendicular to the field in the dipole magnets, then this component of the spin will rotate as the particle moves around the ring. The number of rotations of the spin in one turn of the ring is known as the *spin tune* and is given by

$$\nu_s = \gamma G, \tag{6.5}$$

where γ is the relativistic factor of the electron and G is a physical constant, the anomalous magnetic moment. For electrons, $G \approx 1.159 \times 10^{-3}$. Since the spin tune is dependent on the energy of the particle, the energy spread of particles in a storage ring means that there will be a spread in the spin tunes of the particles. As a result, if a bunch has an initial spin polarisation perpendicular to the field in the dipole magnets, it can quickly become depolarised. However, if the spin polarisation is parallel to the field in the dipole magnets then, neglecting the fields in the quadrupole magnets, the polarisation will be maintained. If we take into account the fields in the quadrupole magnets, then at certain energies (when the spin tune is an integer, or a half integer) there can be resonance effects leading to rapid depolarisation [22].

To understand how spin resonances occur, consider as an example a particle performing vertical betatron oscillations. Each time this particle passes through a quadrupole magnet at a particular point in the ring, the spin vector of the particle will be rotated around a horizontal axis. If the betatron tune is close to an integer, then on the next turn the spin will undergo the same rotation in the field of the quadrupole magnet as before. However, if the spin tune is close to a half integer, then the precession of the spin in the meantime means that the second spin rotation will essentially cancel the first spin rotation. However, if the spin tune is close to an integer, then the two spin rotations will add up, leading to a steadily increasing displacement of the spin vector from its original orientation. The strongest spin resonances usually occur when the spin tune is an integer, that is when the (electron) beam energy is an integer multiple of 440 MeV.

Spin depolarisation provides a way of making a very precise measurement of the energy of a beam in a storage ring; see, for example, [23] and references therein. The beam must initially have some reasonable level of spin polarisation (usually, as a result of the Sokolov–Ternov effect). A suitable component in the storage ring (for example, a small dipole magnet or a solenoid) is used to provide a magnetic field oscillating sinusoidally at a frequency slightly above $2\pi\nu_s/T_0$, where T_0 is the revolution period for particles in the ring. The frequency is then reduced slowly while the beam lifetime is monitored. When the frequency of oscillation of the magnetic field exactly matches $2\pi\nu_s/T_0$, the resonance between the particle spin precession and the oscillating field leads to rapid depolarisation of the beam, resulting in a sudden drop in the Touschek lifetime. Since the revolution period T_0 is known with a very high level of precision (from the frequency of the field in the RF cavities), the spin tune can be determined also with a high level of precision, and the relation between the spin tune and the particle energy (6.5) allows the beam energy to be determined.

In a collider, it can be important for the physics studies to be able to control the spin polarisation of the colliding beams at the interaction point. This can be achieved through suitable configurations of magnetic fields either distributed around the ring, or placed locally near to the interaction point. There has been a significant amount of effort over the years devoted to the understanding and control of spin polarisation in storage rings: see, for example, [24, 25].

6.2.2 Beam–beam effects

In a collider, particles from one beam experience forces from the electromagnetic fields surrounding the opposing beam. The forces are strongest during the collision of two bunches, but there can also be significant effects from the forces acting between bunches not directly in collision, but that are approaching or leaving the interaction point. In general, the effects from forces between opposing beams in a collider are known as *beam–beam effects* [26, 27]; in the case of bunches not directly in collision, the effects are called *long-range beam–beam effects*.

Consider the case of an electron–positron collider. The electric field around (and within) a positron bunch is directed outwards from the centre of the bunch. The particles in a bunch of electrons carry a negative charge, which means that when they collide with a bunch of positrons, they experience a force acting towards the centre of the positron bunch. The fact that the particles in the bunches are moving means that there will also be magnetic forces acting on the particles during a collision. In the case of an electron bunch colliding with a positron bunch, the force on the electrons from the magnetic field generated by the positron bunch will act away from the centre of the positron bunch. Therefore, there is a tendency for the force from the magnetic field to cancel the force from the electric field. However, the force from the electric field is always larger than the force from the magnetic field. The resultant force on the electrons then acts towards the centre of the positron bunch, with a strength that varies as $1/\gamma^2$, where γ is the relativistic factor corresponding to the beam energy, and we assume that both beams have the same energy.

If the beams do not collide exactly head-on but with some displacement between the bunch centroids, then the trajectory of each bunch will be deflected by the beam–beam force. The effect will be similar to a dipole field error, and will lead to some distortion of the closed orbit around the storage ring; this can usually be corrected reasonably easily, using small dipole (steering) magnets at appropriate locations.

The variation of the strength of the beam–beam force with distance from the centre of a bunch leads to focusing effects. This is because up to some distance corresponding roughly to the transverse beam size, the strength of the force will increase with distance from the centre; at distances larger than the beam size, however, the force will fall off with increasing distance. Close to the centre of the bunch, the variation of the force can be approximated as a linear function of the distance from the centre, so that in the horizontal (x) direction, for example, $F_x \propto x$. This is the same variation as in a quadrupole magnet; beam–beam forces can therefore provide focusing in the same way as quadrupoles, at least for particles with small betatron amplitudes. However, the beam–beam force will be focusing (for an electron–positron collider) in the vertical direction as well as in the horizontal direction. This is different from the case of a quadrupole magnet, which provides focusing in one direction but defocusing in the other direction. Also, in a quadrupole magnet, the focal length has the same magnitude (but opposite sign) in the horizontal and vertical directions; however, the beam–beam forces experienced by particles in one beam depend on the size of the opposing beam. Unless the opposing beam has a perfectly circular cross-section, the focal lengths resulting from the beam–beam interaction will be different in the horizontal and vertical directions. Another difference between focusing from a quadrupole magnet and the focusing from the beam–beam interaction is that the beam–beam force is strongly nonlinear: particles at large betatron amplitude will see a force that varies not linearly with distance from the centre of the opposing bunch, but a force that falls off as $1/x^2$ in the horizontal direction, or $1/y^2$ in the vertical direction. The nonlinearity of the force means that particles with different betatron amplitudes will experience different focusing strengths from collisions with the opposing beam.

The focusing strength in a storage ring determines the Courant–Snyder parameters and the betatron tunes. The beam–beam forces will lead to some change in both the Courant–Snyder parameters and the betatron tunes for both beams; however, the nonlinear nature of the beam–beam force means that the changes in these quantities will vary depending on the betatron amplitude of the particle experiencing the force. This leads to a range of betatron tunes, or a *tune spread*, for the particles in a collider, which is characterised by the *beam–beam parameter*. In the case of the vertical motion, the beam–beam parameter is given by

$$\xi_y = \frac{\beta_y^*}{4\pi} \frac{2N_b r_e}{\gamma \sigma_y (\sigma_x + \sigma_y)}, \tag{6.6}$$

where β_y^* is the vertical beta function at the interaction point, N_b is the number of particles in the opposing bunch, r_e is the classical radius of the electron, and σ_x and σ_y are (respectively) the horizontal and vertical beam sizes of the opposing bunch. Usually, in an electron (or positron) storage ring, the vertical beam size is smaller than the horizontal beam size. This means that the vertical beam–beam parameter is normally much larger than the horizontal beam–beam parameter.

If the beam–beam parameter is sufficiently large, then the tune spread can lead to particles crossing betatron resonances, with potentially damaging effects (including the loss of particles from the beam). From that point of view, it is desirable to try to keep the beam–beam parameter as small as possible. However, the beam–beam parameter is related to one of the main figures of merit for a collider, the luminosity $\mathscr{L}$. The luminosity [28] is a measure of the rate of collisions between particles in the opposing beams, and is given by

$$\mathscr{L} = \frac{N_b^+ N_b^- f_0}{2\pi \Sigma_x \Sigma_y}, \tag{6.7}$$

where N_b^+ and N_b^- are (respectively) the number of particles in the positron and electron bunches, f_0 is the rate at which bunches collide, and Σ_x and Σ_y are the sums in quadrature of the horizontal and vertical beam sizes of the colliding beams at the interaction point. The rate of production of new particles in a collider depends on the luminosity: the physics studies, therefore, demand a luminosity that is as large as possible. However, for a given bunch collision rate f_0 (which is limited by the spacing between bunches in the storage rings), an increase in the luminosity requires an increase in the number of particles in the bunches, or a reduction in the beam sizes at the interaction point. Either measure will lead to an increase in the beam–beam parameter. Operation of a collider is often a compromise between achieving a high luminosity, and keeping the beam–beam parameter within a reasonable limit. The practical limit on the beam–beam parameter depends on many factors, including the degree to which quadrupole and higher-order multipole field errors are corrected. Correcting field errors around the machine can help to reduce the strengths of betatron resonances. The better the tuning (error correction) in a collider, the larger the value that can be tolerated for the beam–beam parameter, and the higher the luminosity that can be achieved.

The complex nonlinear and dynamic nature of the forces involved means that reliable estimates of the impact of beam–beam effects usually requires computer modelling. This will be necessary to take into account specific details of the collision, which may include asymmetries between bunch charges and energies, a non-zero crossing angle between the trajectories of the colliding bunches (which is often needed to allow rapid separation of the bunches following the collision, to minimise long-range beam–beam forces) and variation of the beam size along the length of a bunch during collision (that can result from the very small values of the beta functions, needed to achieve small beam sizes). However, computer modelling of beam–beam effects can be challenging for a number of reasons. First, the particle density within a bunch is rarely well-approximated by a simple Gaussian, and often changes during the course of a collision—calculating the fields within and around a bunch during a collision can be computationally expensive, but must be done multiple times for each collision. Second, the effects of the beam–beam forces depend on the optics throughout the storage ring, not just at the interaction point. It is therefore necessary to track particles around the ring in addition to modelling the dynamics during each collision, and in large storage rings particle tracking can be very time-consuming. Third, the modelling should, in principle, be fully self-consistent, so that the forces generated by each beam are calculated from the particle distributions within the colliding bunches, with the distributions evolving properly in response to the forces that the particles experience. Given the complexity of the overall system, it is common to make certain approximations or simplifications; however, these must be applied with care, to avoid undermining the reliability of the results. Computer codes used for modelling beam–beam effects need to be rigorously validated against experimental data, before being applied to predict the impact of beam–beam effects in new operational scenarios, or in future colliders.

6.2.3 Electron-cloud effects

In section 5.2 we discussed the effects that may result from ions becoming 'trapped' in a beam of electrons in a storage ring. The ions may be generated from interactions between residual gas molecules and the beam, or from synchrotron radiation, and may cover a wide range of different masses and electric charges from individual atoms or molecules carrying a single positive electric charge, up to dust particles with very large mass and charge. The free electrons produced by ionisation of an individual atom or molecule are generally expelled from the path of the electron beam by the electrostatic repulsion between the electrons produced by the ionisation and the electrons in the beam. However, in the case of positron beams, any electrons produced in the vacuum chamber can be 'trapped' by the positron beam, in a similar manner to the trapping of positive ions in a beam of electrons. Under certain conditions, the density of electrons within the vacuum chamber of a positron storage ring can build up to levels high enough for interactions between the electrons and the positrons to limit the performance of the storage ring. The effects of the interaction, known as *electron-cloud effects* [29, 30] can include an increase in beam size, various beam instabilities, and even beam loss. Electron clouds can affect proton storage rings as well as positron storage rings.

Clouds of electrons can be created in positron storage rings in a number of different ways. While ionisation of residual gas molecules in the vacuum chamber can contribute to the total number of electrons present, it is usually synchrotron radiation effects, and the generation of 'secondary' electrons that dominate the build-up of an electron cloud. When synchrotron radiation hits the wall of a vacuum chamber, electrons can be released by photoemission. However, electrons can also be released from the wall of the vacuum chamber following the impact of an incident electron. The number of electrons, known as *secondary electrons*, that are released depends on the energy of the incident electron, the angle at which it hits the wall, and on the material and surface structure of the vacuum chamber. Electrons striking the chamber wall at very low energy (a few eV) can result in the release of perhaps just one secondary electron, or no secondary electrons at all. However, with increasing energy, the number of secondary electrons can rise, so that an incident electron with an energy of some tens or hundreds of eV can release multiple secondary electrons. At higher energies (a few keV) the number of secondary electrons tends to fall. Although any electrons initially produced in the vacuum chamber usually have low energy, they can gain energy by being accelerated by the fields around a positron beam.

Depending on the beam parameters and the geometry of the vacuum chamber, electrons appearing near the wall of the vacuum chamber can be accelerated by a bunch of positrons to energies sufficient for them to create several secondary electrons if they hit the wall of the vacuum chamber. If they cross the chamber (and hit the opposite wall) in the time between the arrival of one positron bunch and the next, then a number of secondary electrons can be produced just in time for these new electrons to be accelerated, cross the chamber, and hit the wall releasing further secondary electrons. The result, as more positron bunches go past, is a cascade of electrons, the number of which can increase exponentially. This process, known as *multipacting*, can lead to the rapid build-up of electrons in the vacuum chamber at densities high enough to have damaging effects on the positron beam. The situation can be made more severe by the presence of magnetic fields (from dipole magnets, or quadrupole magnets) that can trap electrons, effectively preventing any low-energy electrons from hitting and being re-absorbed by the wall of the vacuum chamber.

A variety of different methods for preventing the build-up of electron cloud in positron or proton storage rings have been investigated. In the case of positron storage rings, the primary means of control is usually to avoid the generation of electrons by photoemission, since it is often the electrons from this process that initiate any multipacting. To prevent photoemission, the synchrotron radiation produced by the positrons must be prevented from hitting the wall of the vacuum chamber. This can be accomplished by providing a 'secondary' chamber either as an antechamber to the main vacuum chamber, or by providing an extraction port positioned downstream of each dipole magnet (i.e. at locations likely to be impacted by synchrotron radiation). However, eliminating synchrotron radiation from the main chamber entirely can be difficult, because of the complexity of the chamber geometry that would be required, and because of the reflection of radiation from the surface of the chamber. Also, electrons can still be generated by ionisation of residual gas molecules. It is therefore important to achieve extremely good vacuum conditions to minimise the number of residual gas molecules. However, it is again not possible to eliminate residual gas from the chamber entirely: to control the build-up of electron cloud it is often necessary to adopt further measures to limit the production of secondary electrons. This can be done by coating the surface of the chamber with a material having a low secondary electron yield. Materials that have been investigated for this purpose include titanium nitride, and some non-evaporable getter (NEG) materials consisting of various alloys of aluminium, zirconium, titanium, vanadium, and iron. The advantage of NEG coatings is that if used appropriately, they can also help to improve the vacuum inside the chamber. Another way of limiting the generation of secondary electrons is by treating the surface to create numerous fine grooves, so that any incident electrons effectively hit the surface at close to normal incidence, even if the trajectory of the electron is at grazing incidence to the larger-scale surface.

Finally, although magnetic fields from dipole magnets and quadrupole magnets can make electron cloud effects worse by trapping electrons within the chamber, it is possible to use solenoid fields to prevent the build-up of electron cloud. Solenoid fields can be generated by passing an electric current through coils of wire wrapped around the vacuum chamber. An electron released from the wall of the vacuum chamber follows a helical trajectory around the solenoidal field lines that brings it back to the chamber wall where it is safely re-absorbed before it can be accelerated to high energy by the positron beam. This technique is only effective in regions of the accelerator that are otherwise free of electromagnetic fields. In a dipole magnet, for example, the effect of the dipole field will completely dominate any effect on the electrons from a (relatively weak) solenoid field. However, if the solenoid windings can be applied over a sufficient portion of the storage ring circumference, then solenoid fields can be effective in suppressing electron cloud effects.

6.3 The future: 'ultimate' storage rings, and beyond

Electron storage rings for modern light sources and colliders operate in parameter regimes far more demanding in terms of intensity, beam quality and stability, than was the case for early synchrotrons. The push towards more ambitious levels of performance has been driven by demands from the scientists who make use of the facilities, either for studies in particle physics (in the case of colliders) or for studies in a wide range of fields from condensed matter physics to structural biology (in the case of light sources). Improvements in performance have been made possible partly by advances in technology, particularly in the electronics and computing power that underlie the beam diagnostics, feedback systems and control systems on which storage ring operation depends. New parameter regimes have also been made accessible by the tremendous progress that has been made in understanding a wide variety of beam physics phenomena. Some areas, notably in nonlinear dynamics and collective instabilities, continue to pose challenges and there is still significant scope for further advances.

As an illustration of the performance improvements that continue to be achieved, it is interesting to consider the natural emittance of electron storage rings used in synchrotron light sources since the first facilities specifically constructed for the generation of synchrotron radiation. Some representative examples are given in table 6.1. Second-generation light sources constructed in the 1970s and 1980s operated with natural emittances of tens or hundreds of nanometres. These were the first facilities specifically designed and constructed for the production of synchrotron radiation. As experimental techniques developed, users began to demand higher-brightness radiation; at the same time, advances in accelerator science, in particular the development of more advanced lattice designs and the use of insertion devices, allowed increases in radiation brightness of several orders of magnitude. The third-generation facilities that began operation in the last decade of the twentieth century and the first decade of the twenty-first century typically achieved emittances of a few nanometres. More recently, storage rings with emittances of a few hundred picometres have started operation. Although it has been known for some time how to design a storage ring to achieve emittances in this regime, it has required significant advances in technology to address the challenges associated with beam stability and control to make operation of these facilities practicable. Design studies for future facilities aim at similar, or more ambitious

Table 6.1. Year of first operation and natural emittance of some storage rings for synchrotron light sources, constructed since the 1970s. In many cases, the storage rings are capable of operating over a range of parameters, and many have been upgraded since initial commissioning. The emittance values should therefore be taken as representative, rather than definitive.

Facility	Year of first operation	Natural emittance (nm)
SRS [32]	1981	1000
NSLS [33]	1984	9
SRS-HBL [34]	1986	100
ESRF [35]	1993	3.8
ALS [36]	1993	2.0
APS [37]	1995	3.1
SPring-8 [38]	1997	2.4
SLS [39]	2001	5.5
AS [40]	2007	7.0
SOLEIL [41]	2007	3.7
DLS [42]	2007	3.2
PETRA-III [43]	2009	1.2
ALBA [44]	2011	4.6
NSLS-II [45]	2014	0.5
MAX IV [46]	2017	0.26

parameters—the 'ultimate' light source would operate at the diffraction limit, where the radiation brightness is dominated by diffraction effects rather than by the emittance of the electron beam [31].

While storage rings have been of tremendous importance for a wide range of scientific research, they do have some intrinsic limitations that have led to the exploration of alternative accelerator technologies. In particular, for high-energy physics, the emission of synchrotron radiation limits the energy that can be reached in a storage ring of a given size: as the beam energy increases, so does the amount of RF power needed to replace the radiation losses (see section 3.1.1). The synchrotron radiation power can be reduced by increasing the ring circumference; but at some point the size of the ring becomes impractical. It is, of course, possible to build hadron colliders in which the energy limit is set by the field strengths that can be achieved in the magnets, rather than synchrotron radiation effects; the highest energy colliders constructed to date are hadron colliders [47]. However, the physics of hadron collisions is much more complex than that of lepton (electron–electron or electron–positron) collisions, which means that for precision studies in high-energy physics, lepton colliders have an advantage over hadron colliders. To overcome the limitations from synchrotron radiation in electron machines, *linear colliders* have been proposed [48–50], in which beams of electrons and positrons are accelerated in straight linacs. Such an approach was used, for example, in the Stanford Linear Collider (SLC [51]). Similarly, light sources based on linacs are now being constructed that allow operation of free-electron lasers (FELs [52–55]) achieving radiation with brightness several orders of magnitude higher than can be reached using a storage ring. Examples are the LCLS [56], SACLA [57] and the European XFEL [58]; see figure 6.1. Although it is possible to build an FEL based on a storage ring, operating an FEL at the highest levels of brightness results in significant degradation of the quality of the electron beam. This means that state-of-the-art facilities are based on the generation of a fresh electron bunch for every radiation pulse, with the electrons being accelerated in a (straight) linac to the energy required for the FEL and then being dumped after a single shot.

The advantages of linac-based facilities include, therefore, the possibility of reaching very high beam energy (several hundred GeV, for high-energy physics) or very high radiation brightness (for free-electron lasers). In the case of linac-based FELs, it is also possible to achieve radiation pulse lengths in the femtosecond or attosecond regime [55], compared to the picosecond regime typical in storage rings. This opens up exciting possibilities for time-resolved studies, allowing (for example) the investigation of the molecular dynamics involved in chemical reactions. However, linac-based facilities also have significant limitations. In particular, linacs of hundreds of metres, or even kilometres in length are needed to achieve the required beam energy, meaning that the construction costs of such facilities can be high. Since electron beams must be accelerated to the required energy for each bunch collision or radiation pulse, large amounts of RF power are needed to generate good luminosity or brightness. Linac-based facilities can usually serve only a small number of experimental stations at once, whereas storage rings can often deliver luminosity or synchrotron radiation to many experiments simultaneously. Also, achieving the required beam quality and stability in a linac-based facility can be a considerable challenge, because of the way that a fresh bunch is used for each

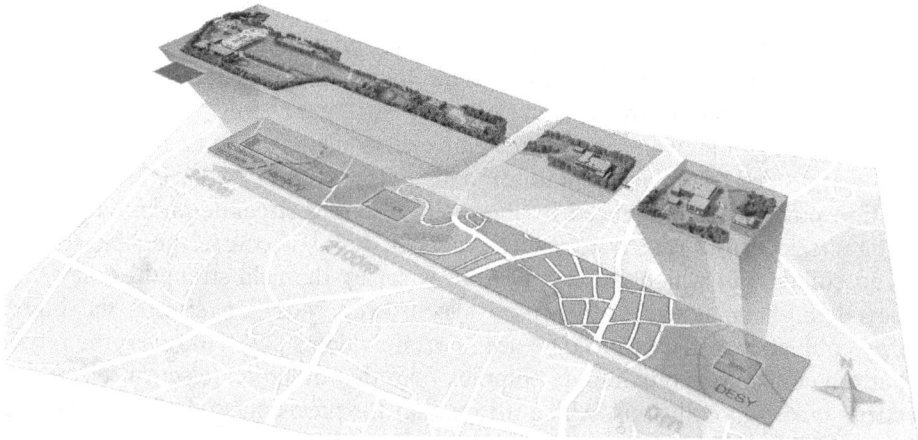

Figure 6.1. Overview of the three sites comprising the European XFEL. The major part of the facility is located in underground tunnels. Three sites above ground provide access to the tunnels. The site DESY-Bahrenfeld marks the beginning of the x-ray laser. From here, the facility is supplied with electrons, energy and cooling. Beneath the site Osdorfer Born (at 2100 m), the beamline splits into a number of separate lines serving different experimental stations. Beneath the site Schenefeld (at 3400 m)—the research campus of the European XFEL facility—the electrons emit the x-ray flashes that are then used for research in the experimental hall. Courtesy of European XFEL (Blunck+Morgen Architekten / tricklabor).

machine pulse. In a storage ring, the beam is maintained in an equilibrium state over long periods, providing a mode of operation that is intrinsically more stable than is possible in a linac, and allowing the use of feedback systems that continuously monitor the beam and provide small adjustments to the various subsystems as required (for example, to control the beam orbit, or to suppress instabilities). The ability of storage rings to serve many users at once is a particular advantage, and despite the particular advantages offered by linac-based facilities means that storage rings are likely to provide the basis for important and highly productive scientific research facilities for some time to come.

References

[1] Goldstein H, Poole C P Jr and Safko J L 2001 *Classical Mechanics* 3rd edn (Boston, MA, USA: Addison-Wesley)

[2] Wolski A 2014 *Beam Dynamics in High Energy Particle Accelerators* (London, UK: Imperial College Press)

[3] Dragt A J 2017 Lie methods for nonlinear dynamics with applications to accelerator physics http://www.physics.umd.edu/dsat/dsatliemethods.html [Online; accessed 3 January 2018].

[4] Forest É 1998 *Beam Dynamics: A New Attitude and Framework* (Amsterdam: Harwood Academic)

[5] Wiedemann H 2015 *Particle Accelerator Physics* 4th edn (New York: Springer)

[6] Yoshida H 1990 Construction of higher order symplectic integrators *Phys. Lett.* A **150** 262–8

[7] Wu Y K, Forest É and Robin D 2003 Explicit symplectic integrator for s-dependent static magnetic field *Phys. Rev.* E **68** 046502

[8] Methodical Accelerator Design http://mad.web.cern.ch/mad/ [Online; accessed 29 December 2017]

[9] Bmad https://www.classe.cornell.edu/bmad/ [Online; accessed 29 December 2017]

[10] elegant http://www.aps.anl.gov/Accelerator_Systems_Division/Accelerator_Operations_Physics/elegant.html [Online; accessed 29 December 2017]

[11] Berz M 2013 Differential algebraic techniques *Handbook of Accelerator Physics and Engineering* 2nd edn ed A Wu Chao, K H Mess, M Tigner and F Zimmermann (Singapore: World Scientific), pp 105–9

[12] Berz M, Makino K and Wan W 2015 *An Introduction to Beam Physics* (Boca Raton, FL: CRC Press)

[13] COSY INFINITY http://www.bt.pa.msu.edu/index_cosy.htm [Online; accessed 29 December 2017]

[14] Forest É, Schmidt F and McIntosh E 2002 Introduction to the Polymorphic Tracking Code http://madx.web.cern.ch/madx/doc/ptc_report_2002.pdf [Online; accessed 3 January 2018]

[15] Forest É 2016 *From Tracking Code to Analysis* (Tokyo, Japan: Springer)

[16] Kleiss R, Schmidt F, Yan Y T and Zimmermann F 1992 On the feasibility of tracking with differential-algebra maps in long-term stability studies for large hadron colliders *Technical Report CERN SL/92–02 (AP), DESY HERA 92 01, SSCL Report SSCL 564* (Geneva, Switzerland: CERN)

[17] Ford G W and Mullin C J 1957 Scattering of polarized Dirac particles on electrons *Phys. Rev.* **108** 477–81

[18] Lee T-Y, Choi J and Kang H S 2005 Simple determination of Touschek and beam-gas scattering lifetimes from a measured beam lifetime *Nucl. Instrum. Methods Phys. Res. Sect. A* **554** 85–91

[19] Sokolov A A and Ternov I M 1964 On polarization and spin effects in the theory of synchrotron radiation *Sov. Phys.: Doklady* **8** 1203–5

[20] Thomas L H 1927 The kinematics of an electron with an axis *Philos. Mag.* **3** 1–22

[21] Bargmann V, Michel L and Telegdi V L 1959 Precession of the polarization of particles moving in a homogeneous electromagnetic field *Phys. Rev. Lett.* **2** 435–6

[22] Lee S Y 1997 *Spin Dynamics and Snakes in Storage Rings* (Singapore: World Scientific)

[23] Wootton K P *et al* 2013 Storage ring lattice calibration using resonant spin depolarization *Phys. Rev. Spec. Top. - Accel. Beams* **16** 074001

[24] Barber D P, Heinemann K and Ripken G 1999 Notes on spin dynamics in storage rings (second revision) *Technical Report DESY M 92-04* (Hamburg, Germany: DESY)

[25] Barber D P 1999 Electron and positron spin polarisation in storage rings—an introduction *Advanced ICFA Beam Dynamics Workshop on Quantum Aspects of Beam Physics* ed P Chen (Singapore: World Scientific), pp 64–90

[26] Herr W and Pieloni T 2014 Beam-beam effects *Proceedings of the CERN Accelerator School 2013: Advanced Accelerator Physics (Trondheim, Norway, 18–29 August 2013)* ed W Herr (Geneva, Switzerland: CERN), 431–60 number CERN-2014-009 (arXiv:1601.05235)

[27] Hirata K 2013 Beam-beam effects in storage ring colliders *Handbook of Accelerator Physics and Engineering* 2nd edn ed A Wu Chao, K H Mess, M Tigner and F Zimmermann (Singapore: World Scientific), pp 169–74

[28] Furman M A and Zisman M S 2013 Luminosity *Handbook of Accelerator Physics and Engineering* 2nd edn ed A Wu Chao, K H Mess, M Tigner and F Zimmermann (Singapore: World Scientific), pp 311–8

[29] Furman M A 2013 Electron cloud effect *Handbook of Accelerator Physics and Engineering* 2nd edn ed A Wu Chao, K H Mess, M Tigner and F Zimmermann (Singapore: World Scientific), pp 163–7

[30] Furman M A 2012 Electron cloud effects in accelerators *ECLOUD12: Joint INFN-CERN-EuCARD-AccNet Workshop on Electron-cloud effects* ed R Cimino, G Rumolo and F Zimmermann (Geneva, Switzerland: CERN), pp 1–8

[31] Borland M 2013 Progress towards an ultimate storage ring light source *J. Phys.: Conf. Ser.* **425** 042016

[32] SRS Accelerator Group 1981 Report on the Daresbury Synchrotron Radiation Source *IEEE Trans. Nucl. Sci.* **NS-28**(3) 2724–8.

[33] National Synchrotron Light Source https://www.bnl.gov/ps/nsls30/ [Online; accessed 3 January 2018]

[34] Thompson D J and Suller V P 1989 Conversion of the SRS to a higher brilliance lattice *Rev. Sci. Instrum.* **60** 1377

[35] Filhol J M 1994 Status of the ESRF *Proceedings of the Fourth European Particle Accelerator Conference (London, UK)* pp 8–12

[36] Jackson A 1993 Commissioning and performance of the Advanced Light Source *Proc. 1993 Particle Accelerator Conference (Washington DC, USA)* pp 1432–5

[37] Galayda J N 1995 The Advance Photon Source *Proc. 1995 Particle Accelerator Conference (Dallas, TX, USA)* pp 4–8

[38] Date S *et al* 1999 Operation and performance of the SPring-8 storage ring *Proc. 1999 Particle Accelerator Conference (New York)* pp 2346–8

[39] Streun A *et al* 2001 Commissioning of the Swiss Light Source *Proc. 2001 Particle Accelerator Conference (Chicago, IL, USA)* pp 224–6

[40] Jackson A 2007 The Australian Synchrotron Project *Proc. 2007 Particle Accelerator Conference (Albuquerque, NM, USA)* pp 911–3

[41] Nadji A *et al* 2007 Commissioning and performance of SOLEIL In *Proceedings of 2007 Particle Accelerator Conference (Albuquerque, NM, USA)* pp 932–4

[42] Bartolini R 2007 The commissioning of the Diamond storage ring *Proc. 2007 Particle Accelerator Conference (Albuquerque, NM, USA)* pp 1109–11

[43] Balewski K 2006 Progress report on PETRA-III *Proc. Tenth European Particle Accelerator Conference (Edinburgh, UK)* pp 3317–9

[44] Einfeld D 2011 ALBA synchrotron light source commissioning *Proc. Second Int. Particle Accelerator Conference (San Sebastian, Spain)* pp 1–5

[45] Willeke F 2015 Commissioning of NSLS-II *Proc. 2015 International Particle Accelerator Conference (Richmond, VA, USA)* pp 11–6

[46] Leemann S C, Sjöström M and Andersson Å 2017 First optics and beam dynamics studies on the MAX IV 3 GeV storage ring *Proc. Eighth International Particle Accelerator Conference (Copenhagen, Denmark)* pp 2756–9

[47] The Large Hadron Collider https://home.cern/topics/large-hadron-collider [Online; accessed 3 January 2018]

[48] Tigner M 1965 A possible apparatus for electron clashing-beam experiments *Il Nuovo Cimento* **37** 1228–31

[49] Aicheler M *et al* 2012 A multi-TeV linear collider based on CLIC technology: CLIC conceptual design report *Technical Report CERN–2012–007, SLAC–R–985, KEK Report 2012–1, PSI–12–01, JAI–2012–001* (Geneva, Switzerland: CERN)

[50] Behnke T *et al* 2013 The International Linear Collider: technical design report *Technical report*, The Linear Collider Collaboration.

[51] Stanford Linear Collider http://www-sldnt.slac.stanford.edu/alr/slc.htm [Online; accessed 2 January 2018]

[52] Freund H P and Antonsen T M Jr 1996 *Principles of Free-electron Lasers* 2nd edn (London, UK: Chapman and Hall)

[53] Saldin E L, Schneidmiller E A and Yurkov M V 2008 *The Physics of Free Electron Lasers* (Berlin, Germany: Springer)

[54] Szarmes E B 2014 *Classical Theory of Free-electron Lasers* (San Rafael, California, USA: Morgan & Claypool Publishers) (Bristol, UK: IOP Concise Physics, IOP Publishing)

[55] Seddon E A *et al* 2017 Short-wavelength free-electron laser sources and science: a review *Rep. Prog. Phys.* **80** 115901

[56] Linac Coherent Light Source https://lcls.slac.stanford.edu [Online; accessed 3 January 2018]

[57] SACLA http://www.riken.jp/en/research/environment/sacla/ [Online; accessed 3 Janaury 2018]

[58] European XFEL https://www.xfel.eu [Online; accessed 3 Janaury 2018]